AF679127

Unified Energy Dynamics

SANDEEP CHAVAN

Published by SANDEEP CHAVAN, 2025.

UNIFIED ENERGY DYNAMICS

First edition. January 11, 2025.

ISBN: 979-8227272379

Written by SANDEEP CHAVAN.

Table of Contents

To the brilliant **scientists** across the world, whose relentless pursuit of knowledge has illuminated the mysteries of the universe.

To **Satyendra Nath Bose** and **Albert Einstein**, whose pioneering work on quantum mechanics and the Bose-Einstein Condensate laid the foundation for understanding the unity and coherence of energy.

To the **philosophical pioneers**, whose timeless wisdom bridges the realms of science and metaphysics, inspiring us to seek deeper truths about existence.

And to my father, **Late Shri Jagannath Chavan**, whose guidance, love, and unwavering belief in me have been the cornerstone of my journey.

This work is a tribute to your vision, brilliance, and legacy.

"In the infinite interface of stability and dynamism, energy is the eternal connector, shaping the cosmos, guiding its evolution, and connecting all that exists. Unified Energy Dynamics is our step toward understanding this universal rhythm."— **Sandeep Chavan**

Preface

The universe, in all its vastness and complexity, has always been a source of wonder and intrigue. From the smallest quantum particles to the immense structures of galaxies, it invites us to ask profound questions: What is the essence of existence? How does energy shape the cosmos? What is our place within this infinite system? These questions, timeless and universal, have driven both scientific inquiry and philosophical reflection throughout human history.

Unified Energy Dynamics (UED) is an attempt to bridge the worlds of science and philosophy, offering a cohesive framework that unifies the principles governing existence. It challenges traditional perspectives by introducing the concept of stable and active energy as the foundation of reality. In doing so, it provides fresh insights into phenomena such as the role of black holes, the flow of time, the behavior of photons, and the interplay of stability and dynamism that sustains the universe's harmony.

This book is not merely a theoretical exploration; it is a call to rethink our understanding of the cosmos. It invites readers to journey beyond established paradigms and consider new possibilities that integrate modern physics, ancient wisdom, and metaphysical inquiry. By doing so, *Unified Energy Dynamics* aspires to illuminate the interconnectedness of all things and inspire a renewed sense of curiosity and purpose.

The inspiration for this work comes from a lifetime of exploration and teaching. My experiences as an educator, coupled with my background in engineering and an enduring fascination with the universe, have shaped the ideas presented here. The questions raised by my students, the discoveries of great scientists, and the timeless wisdom preserved in philosophical texts have all contributed to the formation of this framework.

This book is written for seekers—those who are curious about the mechanics of the universe, the meaning of existence, and the deeper truths that connect us all. It is for scientists, philosophers, students, and anyone who dares to ask "why" and "how." It is for those who believe that understanding the universe is not just about discovering answers but about embracing the journey of inquiry.

I hope this work will inspire you to see the universe not as a collection of isolated phenomena but as a unified, dynamic system. I hope it will challenge you to question, explore, and imagine new possibilities. Most importantly, I hope it will awaken a sense of wonder and interconnectedness, reminding us that we are all part of something far greater than ourselves.

Thank you for joining me on this journey of discovery. May this book spark in you the same sense of awe and curiosity that has guided its creation.

With gratitude and inspiration,

Sandeep Chavan

(Author of Unified Energy Dynamics)

Acknowledgments

This book is the culmination of countless influences, contributions, and inspirations, drawn from humanity's collective pursuit to understand the universe. It stands as a tribute to the interconnectedness of knowledge across disciplines, cultures, and generations.

Gratitude to Scientific Pioneers

I extend my deepest gratitude to the brilliant minds whose groundbreaking work has laid the foundation for **Unified Energy Dynamics (UED)**. Their contributions illuminate the path toward understanding the cosmos:

Satyendra Nath Bose: For his pioneering work on quantum mechanics and Bose-Einstein statistics, providing profound insights into the collective behavior of particles and the principles underlying BECs, a cornerstone of UED.

Isaac Newton: For unveiling the fundamental laws of motion and gravity, offering insights into the balance of stability and motion in the natural world.

Albert Einstein: For redefining space, time, and energy through relativity, inspiring deeper exploration of the infinite energy pool.

Max Planck: For introducing the quantum nature of energy, reshaping our understanding of matter and light.

Niels Bohr: For advancing atomic theory, energy transitions, and the interplay of quantum states.

Stephen Hawking: For his profound work on black holes, entropy, and the nature of time, which profoundly influenced the ideas in this book.

Their discoveries form the bridge between the tangible and intangible, inspiring us to see the universe not as fragmented pieces, but as a unified whole.

Homage to Teachers and Storytellers

I offer heartfelt recognition to the teachers, sages, and storytellers who have preserved and transmitted humanity's philosophical and spiritual heritage. Their dedication ensures that ancient wisdom continues to guide us:

Philosophers of the Gita and Vedanta: The timeless teachings of the Bhagavad Gita, Upanishads, and Vedanta have profoundly shaped humanity's understanding of existence, balance, and purpose. Their emphasis on unity, cycles of creation and dissolution, and the interplay of stability and dynamism resonates deeply with UED.

Scribes and Preservers: To the countless unnamed individuals who safeguarded sacred texts, ensuring their survival through time.

Teachers and Gurus: For making wisdom accessible through storytelling, teaching, and reflection, inspiring seekers to explore both inner and outer universes.

Their contributions remind us of the enduring relevance of philosophical inquiry in understanding the cosmos.

Appreciation for My Students

To my students—your curiosity and relentless questioning have been my greatest source of inspiration. Your willingness to engage with complex ideas and challenge norms has been instrumental in shaping the concepts within this book. You remind me that the pursuit of knowledge is a dynamic and collective endeavor, and for that, I am deeply grateful.

Personal Gratitude

To my family, mentors, and friends—thank you for your unwavering support, encouragement, and belief in this project. Your insights, patience, and feedback have been invaluable throughout this journey.

The Universe as the Ultimate Teacher

Finally, I acknowledge the universe itself—a boundless source of inspiration and wonder. Its harmony, vastness, and intelligence serve as the ultimate guiding principles for this work. **Unified Energy Dynamics** is a humble attempt to honor the intricate beauty of existence and to share its profound lessons.

Closing Words

To all who have contributed, preserved, and inspired the pursuit of knowledge—whether through scientific discovery, philosophical inquiry, or the preservation of timeless wisdom—this book is dedicated to your efforts. Thank you for enriching humanity's understanding of the cosmos and for being a part of this collective journey.

Sandeep Chavan
Author

Part I: Foundations of Energy and Existence

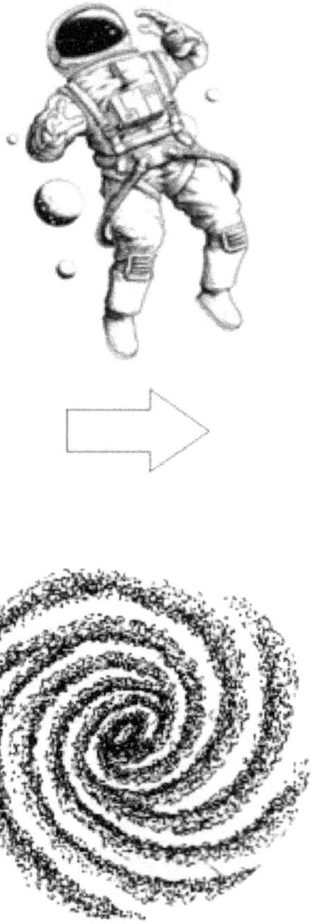

Chapter 1. Introduction: Rediscovering the Universe's Essence

Humanity has long been captivated by the mysteries of existence, driven by an insatiable curiosity to uncover the underlying principles that govern the cosmos. From ancient philosophies to cutting-edge scientific discoveries, this quest has spanned centuries, weaving together threads of observation, imagination, and reason. Yet, despite remarkable strides in understanding, the picture remains incomplete—a mosaic of fragmented theories, isolated disciplines, and unanswered questions.

At the heart of this pursuit lies the yearning for a unified framework, one that seamlessly integrates the macroscopic and microscopic, the tangible and intangible, the physical and metaphysical. The need for such unification is not merely academic; it reflects humanity's intrinsic desire to comprehend its place within the grand tapestry of existence.

1.1 Revisiting the Need for a Unified Understanding

Throughout history, humanity has developed frameworks to explain the natural world. These frameworks, rooted in science and philosophy, have brought us profound insights into the mechanics of existence. However, they often operate in isolation, addressing specific aspects of reality while leaving others unexplained. This fragmentation, though necessary for specialization, creates gaps in our understanding of the universe as a unified whole.

The Strengths and Limitations of Current Frameworks

Newtonian mechanics revolutionized our understanding of motion, providing a precise and predictive model for the behavior of objects on Earth and in the cosmos. Its principles form the bedrock of classical physics, enabling technological advancements and shaping our understanding of forces. Yet, Newton's laws falter when faced with phenomena beyond terrestrial scales, such as the curvature of spacetime or the probabilistic behaviors of particles.

Einstein's General Theory of Relativity advanced our understanding of gravity, presenting it not as a force but as the curvature of spacetime caused by mass and energy. This framework elegantly describes the motion of planets, the warping of light around massive objects, and the expansion of the universe. However, relativity struggles to reconcile with quantum mechanics, the dominant theory governing the behavior of particles at the subatomic level.

Quantum mechanics, in turn, reveals a reality governed by probabilities, where particles can exist in superposition, entangle across vast distances, and exhibit wave-like behaviors. While immensely powerful in explaining microscopic phenomena, quantum mechanics remains detached from the macroscopic domain of relativity, leaving unanswered questions about how the two realms connect.

The Fragmentation of Knowledge

These scientific paradigms, while individually brilliant, function like isolated islands of knowledge. Newton's mechanics governs the everyday world; relativity explains the cosmic scale; and quantum mechanics deciphers the subatomic. Yet, they do not converge into a cohesive framework. This fragmentation leaves critical questions unanswered, such as:

- ✓ How do the principles governing galaxies relate to those governing atoms?
- ✓ What connects the force of gravity with the other fundamental forces of nature?
- ✓ Is there a common origin for the laws that describe both the very large and the very small?

Such disconnection highlights the pressing need for a unifying framework, one that can integrate the strengths of these paradigms while addressing their limitations.

The Universe as a Unified Entity

The universe is not a collection of independent systems operating in isolation; it is a singular, interconnected entity. The forces that shape galaxies also influence atoms. The principles governing the motion of planets and stars emerge from the same fabric that gives rise to the quantum behaviors of particles. To truly understand this reality, we must move beyond fragmented theories and embrace a unified approach that transcends disciplinary boundaries.

A Call for Integration

A unified understanding must do more than reconcile existing theories; it must provide a framework that explains the underlying principles connecting all phenomena. Such a framework would allow us to view the universe not as a collection of separate parts but as a harmonious whole. It would bridge the gap between science and philosophy, combining empirical rigor with profound reflection to address questions about existence, consciousness, and the nature of reality.

Unified Energy Dynamics: A Step Forward

This book introduces Unified Energy Dynamics (UED) as a step toward this integration. By recognizing energy as the foundational essence of all that exists, UED seeks to bridge the divides between classical physics, relativity, and quantum mechanics. It proposes that energy, in its dual forms—stable and active—serves as the common thread linking all phenomena, from the motions of planets to the behavior of particles.

This perspective challenges us to rethink traditional boundaries, to dissolve the silos of knowledge that have constrained our understanding, and to explore a vision of the universe that is unified, interconnected, and profoundly coherent. Through this lens, we may not only deepen our comprehension of reality but also uncover the principles that govern the cosmos in its entirety.

1.2 Energy as the Essence of Reality

At the heart of the universe lies energy—a concept both deeply familiar and profoundly mysterious. Often understood as a measurable physical quantity, energy is, in fact, the fundamental essence of existence. It is not confined to the realm of equations or scientific observations; rather, it is the invisible thread that weaves together the cosmos, transcending the boundaries between the physical and the metaphysical. From the radiant energy of stars to the delicate impulses of human thought, energy is the ultimate force driving all motion, transformation, and continuity.

Energy beyond Measurement: A Universal Principle

Energy is frequently quantified in terms of heat, work, or motion, defined by specific parameters within physics. However, such definitions capture only a fraction of its true nature. Energy is not merely a phenomenon to be measured; it is the foundation upon which all phenomena rest. It is the dynamic interplay that animates the cosmos, a silent yet omnipresent force that fuels every process and structure, from the smallest subatomic interactions to the vast galactic movements.

In this light, energy becomes more than a scientific construct—it is the cornerstone of existence itself. It permeates the fabric of space and time, bridging the tangible with the intangible and linking the observable with the imperceptible. This broader perspective allows us to explore energy not just as a force or quantity but as the essence of reality.

The Dual States of Energy: Stable and Active

This book introduces a transformative understanding of energy as existing in dual states: stable and active. These two states are not opposing but complementary, forming a dynamic relationship that underpins all phenomena.

Stable Energy: This is the silent, unchanging reservoir of existence. Stable energy is the latent essence from which all forms emerge and into which they dissolve. It is the foundation of continuity, providing the stability required for the cosmos to maintain coherence. Without stable energy, the universe would lack the structural integrity necessary to sustain its complexity.

Active Energy: In contrast, active energy is the force of motion and transformation. It drives the processes that create, sustain, and evolve the observable universe. Active energy manifests in phenomena such as light, heat, gravitational waves, and the interactions that shape matter and life. It is the dynamic expression of change, ensuring the perpetual renewal and evolution of the cosmos.

These two states are intrinsically linked, forming a cycle where active energy arises from the stable reservoir, powers the universe's processes, and ultimately returns to its stable state. This interplay creates a harmonious balance, sustaining the infinite dynamics of existence.

Energy as the Architect of the Cosmos

Energy, in its dual states, acts as the architect of the universe. It shapes the fabric of space, dictates the flow of time, and governs the interactions of matter. From the birth of stars to the dance of subatomic particles, energy is the driving force behind every process:

Creation and Transformation: Active energy brings matter into existence, fuels the fusion reactions in stars, and propels the expansion of the universe. It drives the cycles of life and death, the rise and fall of galaxies, and the endless transformations that define the cosmos.

Equilibrium and Continuity: Stable energy ensures that these transformations occur within a coherent framework. It anchors the universe, preventing chaos and enabling the emergence of order and complexity.

The Infinite Interplay: Together, stable and active energy create a universe that is both dynamic and stable, infinite yet interconnected. This interplay ensures that the cosmos remains vibrant and evolving, while simultaneously grounded in an unchanging essence.

Energy as the Bridge Between Science and Experience

Energy's influence extends beyond the physical realm. It connects the measurable phenomena of science with the experiential realities of life and consciousness. The radiance of light, the pull of gravity, the flow of thought—all are expressions of energy's dynamic states. This universality makes energy the bridge between disciplines, uniting the empirical rigor of science with the introspective depth of philosophy and metaphysics.

This perspective not only enriches our understanding of energy but also reshapes our view of existence. It reveals a cosmos that is not fragmented but unified, where every phenomenon, from the smallest quantum fluctuation to the largest cosmic structure, is part of an interconnected whole powered by the dynamics of energy.

A New Paradigm for Understanding Reality

By recognizing energy as the essence of reality, we gain a framework that transcends traditional boundaries. This perspective invites us to rethink long-held assumptions about the universe's workings and embrace a holistic understanding of existence. It challenges us to view energy not just as a force or resource but as the fundamental fabric of all that is.

In this unified framework, energy is both the silent source and the active expression of reality—a principle that binds together science, philosophy, and metaphysics into a cohesive narrative. This book embarks on a journey to explore energy's profound role, unveiling its essence as the cornerstone of the cosmos and the key to understanding the interconnectedness of all things.

1.3 Bridging Science, Philosophy, and Metaphysics

The pursuit of understanding the universe demands more than technical precision; it requires a holistic approach that unites diverse ways of knowing. Science, philosophy, and metaphysics, each with their distinct methodologies and inquiries, have traditionally been treated as separate disciplines. However, their integration offers a more profound comprehension of reality—one that transcends the limitations of any single perspective.

Science: The Lens of Observation and Measurement

Science has been humanity's most powerful tool for exploring the tangible universe. Through observation, experimentation, and mathematical modeling, it has unraveled the mechanics of planetary motion, the nature of light, and the quantum behavior of particles. Science gives us the means to quantify phenomena, revealing the intricate details of the physical world. Yet, despite its precision, science often stops short of addressing the "why" behind the "how." It measures the cosmos but struggles to grasp its ultimate essence.

Philosophy: The Framework of Interpretation

Where science provides the data, philosophy interprets it, offering frameworks to understand meaning and purpose. Philosophy encourages us to ask questions that lie beyond empirical reach: What is the nature of existence? What does it mean to be conscious in a universe governed by laws of physics? These inquiries challenge the assumptions of science, opening pathways to deeper truths. While philosophy lacks the empirical tools of science, its introspective nature offers insights that are equally vital in the quest for a unified understanding.

Metaphysics: Exploring the Transcendent

Metaphysics ventures into the realms that defy direct observation, probing the fundamental nature of reality, time, space, and existence itself. It seeks to understand what lies beneath the observable universe, exploring concepts like causality, infinity, and the ultimate source of all that is. While often dismissed by strict empiricists, metaphysics complements science and philosophy by addressing the profound mysteries that remain unsolved.

Unified Energy Dynamics: A Bridge across Realms

Unified Energy Dynamics (UED) emerges as a framework that connects these disciplines, grounding them in the universal principle of energy. By understanding energy as the essence of reality, UED provides a common language for science, philosophy, and metaphysics. This approach invites collaboration rather than division:

- ✓ **For Science:** UED extends the boundaries of empirical inquiry, connecting the micro-level phenomena of quantum mechanics with the macro-scale structures of the cosmos.

✓ **For Philosophy:** It offers a foundation for interpreting the interconnectedness of existence, grounding abstract concepts in the tangible dynamics of energy.

✓ **For Metaphysics:** UED provides a framework to explore questions about the origins and nature of reality, bridging the measurable with the immeasurable.

Reimagining the Universe through Energy

Energy, in its dual states of stability and activity, serves as the universal thread that connects all phenomena. It unites the seemingly disparate realms of matter and mind, the physical and the intangible. Through this lens, light, gravity, time, and even consciousness are seen not as isolated phenomena but as expressions of energy's dynamic interplay.

This perspective reimagines the universe as a cohesive whole, dissolving the artificial boundaries between science, philosophy, and metaphysics. It challenges us to view reality as a unified system where every phenomenon is interconnected, each influencing and being influenced by the others.

Inviting New Questions and Expanding Understanding

By bridging these disciplines, UED not only synthesizes existing knowledge but also fosters new lines of inquiry. What role does consciousness play in the energy dynamics of the universe? How do stable and active energy states shape the emergence of life and the evolution of galaxies? What principles underlie the transitions between energy states, and how do they govern the flow of existence? These questions invite us to explore the cosmos with renewed curiosity, pushing the boundaries of what we know.

A Framework for Discovery

Unified Energy Dynamics stands as an invitation to rethink our approach to understanding the universe. It embraces the precision of science, the interpretive depth of philosophy, and the transcendent insights of metaphysics, weaving them into a cohesive narrative. This framework does not claim to provide all the answers but rather offers a starting point for an ever-evolving exploration of existence.

In bridging science, philosophy, and metaphysics, we take a step closer to unraveling the mysteries of the cosmos, recognizing the universe not as a collection of isolated parts but as a unified whole—boundless, dynamic, and profoundly interconnected.

1.4 A Call to Rediscover

This introduction extends an invitation to embark on a transformative journey—a journey to reimagine the universe through a lens that unites, rather than divides. By revisiting the essence of existence, we aim to dissolve the artificial barriers between science, philosophy, and metaphysics, unveiling a vision of reality that is cohesive, harmonious, and profoundly interconnected.

At the heart of this exploration lies energy—both in its silent stability and dynamic action. Energy is not merely a concept confined to physics; it is the universal language that defines existence, connecting the tangible and the intangible, the measurable and the mysterious. It is the thread that binds the vast cosmos with the intricate dimensions of consciousness, offering a perspective that is as expansive as the universe itself.

Through the chapters that follow, we will delve into this framework, uncovering the principles that govern the physical universe while illuminating the deeper, intangible dimensions of meaning and awareness. This journey invites us to question, to explore, and to embrace the unity that underpins all phenomena. Together, let us rediscover the universe's essence, guided by the understanding that energy, in its many forms, is the cornerstone of all that is.

Chapter 2: Energy as the Universal Medium

As we delve deeper into the essence of the universe, we encounter a concept both foundational and profound: energy as the universal medium. Expanding on the unified framework introduced earlier, this chapter explores energy not merely as a measurable quantity, but as the silent orchestrator of all existence. Energy is revealed as a dynamic interplay of stability and motion—a duality that underlies every phenomenon in the cosmos. It fuels the brilliance of stars, sculpts the grand architecture of galaxies, and animates the subtleties of consciousness, serving as the unifying thread that connects all that exists.

Here, energy is redefined beyond conventional scientific boundaries, framed through its dual states of stability and dynamism. We examine the timeless principles of conservation, transformation, and entropy, illustrating how energy transitions drive evolution and sustain universal balance. Finally, we explore the interactions within the infinite energy pool—a boundless medium that bridges the micro and macro realms—revealing how its continuous interplay shapes the cosmos through cycles of creation and dissolution.

2.1 Redefining Energy: Stability and Dynamism

Energy, the cornerstone of existence, manifests in two fundamental states: **stable** and **active**. These dual states form the foundation of all phenomena, driving the creation, transformation, and dissolution of the universe. In the framework of **Unified Energy Dynamics (UED)**, energy transcends its conventional definition as a measurable physical quantity and emerges as the essence of reality—a unifying force that bridges the tangible and intangible dimensions of existence.

The Dual States of Energy: Stability and Dynamism

Stable Energy: The Silent Framework

Stable energy is the unchanging bedrock of the cosmos, an infinite reservoir that underlies all creation and existence.

Foundation of Reality: This state of energy acts as the universe's steady framework, providing coherence and balance amidst the dynamic interplay of forces. It remains unaffected by transient disturbances, serving as the universal medium within which all phenomena unfold.

The Energy Medium: Stable energy binds the cosmos, maintaining equilibrium across scales—from the micro-level interactions of particles to the expansive dynamics of galaxies.

Cosmic Reservoir: It is the destination to which all forms of active energy eventually return, preserving the cyclical nature of existence.

Active Energy: The Force of Change

Active energy represents dynamism, the driver of motion, transformation, and all observable phenomena.

Catalyst of Motion: It powers the movements of particles, the orbits of celestial bodies, and the intricate processes of life. From the radiance of stars to the biological energy fueling living organisms, active energy is the observable manifestation of change.

Interaction with Stability: Active energy interacts with the stable pool, generating localized phenomena such as the formation of stars, the emergence of life, and the dissolution of matter in black holes. These events exemplify the perpetual cycle of creation and dissolution.

Consciousness as Energy in Transit

Central to this framework is the understanding of **consciousness** as a dynamic transit between stable and active energy states. Consciousness acts as the bridge that connects the unchanging stability of the energy pool with the dynamic actions of active energy, shaping existence across all scales.

Human Consciousness: Unique and Perceptive

In humans, consciousness emerges through the interplay of active energy (the soul) and its functional manifestation (the mind):

The Soul as Dynamic Awareness: Active energy manifests as the soul, the core of human dynamism and awareness. It is the source of perception, creativity, and intellect.

The Mind as a Functional Bridge: The mind, functioning as the soul's operational interface, enables humans to observe, analyze, and interact meaningfully with the physical world. This unique interplay allows for the development of intellect and the creation of abstract ideas, anchoring humans as conscious participants in the universe.

Consciousness in Matter: Subtle but Present

Even in matter, consciousness exists at a quantum level, reflecting the interaction between stable and active energy:

Quantum Behaviors: Phenomena such as wave-particle duality, superposition, and entanglement demonstrate a form of "awareness" in matter, showcasing its interaction with the energy medium.

Universal Connection: This quantum consciousness links all forms of matter to the stable energy pool, emphasizing the interconnectedness of existence.

The Interplay of Stability and Motion

The duality of energy as stable and active creates a harmonious interplay that governs the universe's dynamics.

Interconnected Dynamics

Active Energy in Daily Life: Confined within matter, active energy drives the observable phenomena we encounter—motion, heat, light, and biological processes.

Stable Energy's Omnipresence: While active energy fuels change, stable energy provides the context within which these changes occur, ensuring equilibrium across systems.

Creation and Dissolution

Formation of Matter: Disturbances within the energy medium trigger transitions from stability to activity, forming matter and initiating cosmic events.

Return to Stability: When the connection to the energy medium is severed—whether through the dissolution of matter or the cessation of biological life—active energy returns to the stable pool, completing the cyclical flow of existence.

Observable Across Scales

Micro-Level Phenomena: At the quantum level, particles exhibit motion and behaviors governed by active energy, reflecting its dynamic presence.

Macro-Level Harmony: Stable energy ensures the coherence of galaxies and cosmic structures, maintaining the universe's order and connectivity.

Energy as the Essence of Creation and Dissolution

Energy, in its dual states, orchestrates the cycle of existence: creation, transformation, and dissolution.

✓ **Active Energy's Role:** It drives motion, powers evolution, and initiates change, giving rise to the observable universe.

✓ **Stable Energy's Role:** It serves as the unchanging reservoir that absorbs disturbances, sustains equilibrium, and provides the foundation for all transformations.

✓ **A Unified Framework:** Together, these states explain the formation of matter, the evolution of systems, and the inevitable return of all phenomena to stability.

A Unified Perspective on Energy

This redefinition of energy as both stability and dynamism bridges scientific and philosophical perspectives, offering a cohesive framework for understanding the universe. By recognizing energy as the universal medium, we connect the micro and macro realms, the physical and the metaphysical, the measurable and the abstract.

Through this lens, the universe emerges not as a fragmented collection of phenomena but as a harmonious interplay of stability and motion—bound together by the eternal dance of energy. It is a vision of existence that is as profound as it is unifying, inviting us to explore the cosmos with a renewed sense of wonder and coherence.

2.2 Conservation, Transformation, and Entropy

Building on the dual nature of energy as stable and active, this section delves into the fundamental principles that govern energy's behavior: **conservation, transformation, and entropy**. These principles form the backbone of energy dynamics, shaping the universe's evolution and providing a framework for understanding its perpetual cycles of creation and dissolution. By exploring how energy is conserved, transformed, and dispersed, we reveal the mechanisms driving change, maintaining equilibrium, and guiding the irreversible flow of existence.

The Principles of Energy Conservation across Systems

At the core of the universe lies an immutable truth: **energy cannot be created or destroyed.** Known as the **law of conservation of energy**, this principle ensures that energy merely transitions between forms while its total quantity remains constant across systems.

Conservation as a Universal Constant

Redistribution, Not Loss: In closed systems, energy transitions seamlessly without loss or gain, merely redistributing among different forms. **For example,** A **pendulum** demonstrates this principle as it oscillates, converting potential energy into kinetic energy and vice versa. Similarly, **nuclear fusion in stars** transforms mass into radiant energy, adhering to conservation on a cosmic scale.

Cosmic Implications: The universe itself operates as an overarching closed system, where stable and active energy interplay to sustain equilibrium. The continuous redistribution of energy ensures that all transformations contribute to the larger coherence of the cosmos.

Conservation across Stable and Active States

Stable Energy as the Reservoir: Stable energy serves as the infinite reservoir, maintaining balance and absorbing disturbances within the universal energy pool.

Active Energy as the Driver: Active energy manifests in dynamic processes, transitioning between motion, heat, light, and other observable phenomena.

Unbroken Continuity: These transitions always adhere to the principle of conservation, maintaining the total energy of the system while driving its processes.

Implications for Universal Equilibrium

Conservation underpins the cosmos's coherence, ensuring that transformations between energy states are balanced.

It allows for **perpetual interaction**, as energy flows between stability and motion, enabling the cycles of creation and dissolution that define the universe's evolution.

How Energy Transitions Drive Transformation and Evolution

Energy transitions act as the engine of transformation, catalyzing the evolution of systems, fostering complexity, and sustaining the dynamic interplay of forces that shape existence.

Energy as the Catalyst of Change

When **stable energy** transitions to **active energy**, it initiates motion, interaction, and creation: **For example: Star formation** results from gravitational compression of gas clouds, transforming potential energy into nuclear fusion. Similarly, **Life processes** emerge from the conversion of chemical energy into biological activity, demonstrating energy's role in existential evolution.

Transformations Across Scales

At the Quantum Level: Particles exchange energy through interactions, manifesting phenomena such as **wave-particle duality** and **quantum entanglement**.

At the Cosmic Level: Gravitational waves propagate through spacetime, reshaping its curvature and influencing celestial bodies.

These transitions highlight the interconnectedness of quantum and cosmic realms, governed by unified principles of energy dynamics.

Guided Evolution

Systems evolve as energy transitions push them toward greater complexity or stability: For instance, **Galaxies** form and mature over billions of years through the redistribution of mass and energy. Similarly, **Biological systems** adapt and grow, driven by the continuous flow of energy through metabolic processes.

This evolution is not random but guided by the intricate balance between stability and motion, fostering coherence amid change.

Entropy: A Measure of Energy Dispersal and System Change

While energy is conserved, its ability to perform useful work diminishes as it transitions between states. **Entropy** quantifies this dispersal, reflecting the inherent tendency of systems to progress toward disorder.

Entropy and Energy Dispersion

Entropy measures the spread of energy within a system, indicating the degree of randomness:

Thermodynamic Processes: Heat naturally flows from regions of high temperature to low temperature, increasing entropy.

Cosmic Phenomena: Energy dispersal through **radiation** and **gravitational waves** contributes to the overall entropy of the universe.

This dispersal ensures energy transitions are directional, establishing the **arrow of time** for living entities and physical processes.

The Role of Entropy in System Evolution

As energy disperses, systems transition from ordered states to less ordered configurations:

A **star's lifecycle** exemplifies this principle, as nuclear fusion depletes its fuel and increases entropy.

On Earth, **ecosystems** harness solar energy to sustain life, yet their processes contribute to the planet's broader entropy.

Entropy, while limiting reversibility, drives transformation, guiding systems toward their eventual states of equilibrium.

Entropy in the Universal Context

The universe, as a dynamic system, is fundamentally governed by the principle of increasing entropy. This relentless rise shapes cosmic evolution, driving expansion and determining the universe's ultimate trajectory.

Expansion and Energy Dispersion

As **galaxies drift apart** due to cosmic expansion, the energy within them disperses across space, increasing the cosmos's entropy.

This spreading ensures the universe's irreversible progression toward greater states of disorder while maintaining its dynamic equilibrium.

Black Holes and Entropy

Black holes, often viewed as endpoints of energy accumulation, paradoxically contribute to universal entropy through **Hawking radiation**:

Energy escaping the event horizon exemplifies quantum-level interactions, dispersing energy back into the cosmos and adhering to entropy's principles.

Perpetual Cycles of Creation and Dissolution

Entropy limits reversibility, preventing energy from spontaneously reconcentrating into its original forms.

However, it also enables the cycles of creation and dissolution, sustaining the dynamic equilibrium of the cosmos where matter forms, evolves, and dissolves back into stability.

Energy, Transformation, and the Flow of Existence

The principles of **conservation**, **transformation**, and **entropy** form the backbone of energy dynamics, offering profound insights into the universe's perpetual motion:

Conservation ensures continuity, allowing energy to flow and interact without loss.

Transformation drives evolution, shaping systems across scales and bridging the quantum and cosmic realms.

Entropy defines directionality, emphasizing the irreversibility of processes and the universe's journey toward equilibrium.

Together, these principles illuminate the universe as an intricate dance of energy—dynamic, evolving, and harmonious. By understanding these forces, we grasp the profound mechanisms that sustain existence and define the perpetual flow of reality.

2.3 Interactions within the Infinite Energy Pool

The **infinite energy pool** serves as the universal medium of existence—a boundless and interconnected reservoir that binds the cosmos together. Far from being an abstract concept, it represents the tangible framework within which all phenomena occur, bridging the realms of stability and motion. Through this medium, energy transitions, interacts, and evolves, shaping reality across every scale. By exploring the interplay between **active energy** and the **stable reservoir**, we uncover the mechanisms that govern the cosmos, from subatomic particles to galaxies, and from physical matter to consciousness itself.

The Energy Pool as the Universal Medium of Existence

The infinite energy pool embodies the totality of energy in the universe, encompassing both **stable energy** (the unchanging reservoir) and **active energy** (the force of transformation). It is the fundamental essence from which all phenomena emerge and into which they eventually dissolve.

Stability as the Backbone

The energy pool's stable reservoir acts as the silent foundation of reality: It ensures that the fundamental laws of physics remain constant across time and space. Stable energy provides the latent potential needed to support the dynamic processes driven by active energy. Without this stability, the cosmos would lack coherence, leading to chaos rather than order and continuity.

Universality of the Energy Pool

Unlike localized phenomena such as stars, planets, or black holes, the energy pool is **universal and infinite**: It permeates every corner of existence, spanning the smallest quantum systems to the vast intergalactic voids. All interactions—whether microscopic or macroscopic—occur within this medium, making it the ultimate unifier of reality.

The Medium for Existence

Everything observed—**matter**, **motion**, **light**, and even **consciousness**—exists within and because of this infinite energy pool. The pool is not merely a backdrop but an **active participant**, shaping and sustaining the interactions that define the universe.

The Interplay between Active Energy and the Stable Reservoir

The dynamic relationship between **active energy** and the **stable reservoir** drives the perpetual motion and transformation of the cosmos. This interplay reveals the delicate balance between creation and stability, shaping both the tangible and intangible aspects of reality.

Active Energy: The Force of Change

Active energy represents motion, interaction, and transformation: It manifests in forms such as **light**, **heat**, **motion**, and **biological processes**, driving observable phenomena. From the nuclear fusion of stars to the chemical reactions sustaining life, active energy fuels creation and evolution. It interacts with the stable reservoir to create **localized disturbances**, forming matter and giving rise to physical reality.

Stable Energy: The Reservoir of Potential

Stable energy serves as the repository of all latent potential: It absorbs disturbances created by active energy, maintaining equilibrium within the infinite energy pool. When active energy dissipates, it reintegrates into the stable reservoir, completing the cycle of creation and dissolution. This stability prevents chaos, ensuring the coherence of the universe despite constant motion.

The Feedback Loop of Energy Dynamics

The interaction between active and stable energy forms a **self-sustaining feedback loop**: Active energy draws from the stable reservoir to fuel motion and transformation. When active energy dissipates or transitions, it returns to the stable pool, preserving equilibrium. This cyclical interplay ensures that the energy pool remains balanced, supporting both dynamic processes and universal stability.

How Energy Dynamics Govern the Cosmos on All Scales

The principles of energy dynamics, rooted in the interaction between active and stable energy, operate across every scale of existence. These dynamics govern the behavior of particles, the formation of celestial bodies, and the evolution of life and consciousness.

Quantum Scale

At the microscopic level, energy dynamics manifest in phenomena such as **wave-particle duality** and **quantum entanglement**: Particles exhibit behaviors influenced by their interactions with the stable energy pool. Active energy drives motion and interaction, while stable energy provides the coherence necessary for quantum systems to function. This constant interplay underpins the behavior of matter and energy at the smallest scales.

Cosmic Scale

On a universal scale, energy dynamics shape the motion of galaxies, the propagation of gravitational waves, and the lifecycle of stars: **Gravitational forces** emerge as interactions between mass and the stable energy medium, defining the architecture of the cosmos. Active energy fuels processes like supernovae and the radiation of light, driving the evolution of celestial bodies. The infinite energy pool ensures that these phenomena remain interconnected, governed by the same principles of stability and transformation.

Consciousness and Life

Even at the level of life and consciousness, energy dynamics are at work: **Consciousness** emerges as the interaction of active energy with stable energy, creating awareness and perception. Biological systems act as **energy loops**, converting active energy into life processes while remaining grounded in the stable reservoir. This connection reveals the universal applicability of energy dynamics, extending beyond the physical to encompass the experiential.

The Infinite Energy Pool as the Foundation of Reality

The infinite energy pool is more than a theoretical construct—it is the essence of existence. Its interactions define the universe's behavior, bridging the microscopic and macroscopic, the physical and metaphysical.

A Unifying Medium: The energy pool connects all phenomena, providing the foundation for a **cohesive and interconnected reality**.

A Source of Balance: Its stable reservoir ensures order and continuity, while its active energy drives motion and change, creating a harmonious interplay.

A Framework for Understanding: By studying the interplay between active and stable energy, we uncover the principles that govern the cosmos, revealing a universe that is **dynamic**, **harmonious**, and **infinite**.

The Dance of Creation and Dissolution

Through the **interactions within the infinite energy pool**, the universe sustains its perpetual cycles of creation, transformation, and dissolution. This boundless medium is not merely a passive backdrop but an active participant, shaping and supporting all phenomena. By understanding this framework, we see the universe not as a fragmented collection of isolated processes but as a single, interconnected system—unified by the eternal principles of energy dynamics.

Summary: Energy as the Universal Medium

This chapter of the UED framework introduces the profound concept of **energy as the universal medium**, emphasizing its dual states and infinite nature. Below is a point-wise summary:

1. Energy as the Foundation of Existence

Definition Beyond Conventional Science: Energy is redefined not merely as a measurable quantity but as the essence of existence, orchestrating the cosmos at every scale. It is portrayed as a unifying thread that bridges the tangible and intangible dimensions of the universe.

Dynamic Interplay: Energy operates as a balance of stability and motion, driving the cycles of creation, transformation, and dissolution.

2. The Dual States of Energy

Stable Energy: Acts as the **unchanging bedrock of the cosmos**, providing a steady framework for existence. Functions as the **universal reservoir**, where active energy dissipates and re-equilibrates. Maintains equilibrium across micro (atomic) and macro (cosmic) scales.

Active Energy: Represents **dynamism and transformation**, fueling observable phenomena like the motion of particles, the radiance of stars, and biological life. Interacts with the stable energy medium to produce localized events, such as star formation and black hole dynamics.

3. Energy Transitions and Universal Evolution

Conservation and Transformation: The chapter revisits the timeless principles of energy conservation and entropy, highlighting how energy transitions sustain universal balance.

Cycles of Creation and Dissolution: The interplay between stable and active energy forms drives cosmic and quantum-scale phenomena, ensuring the perpetual evolution of the universe.

4. Consciousness as Energy in Transit

The Role of Consciousness: Consciousness is described as a **dynamic bridge** between stable and active energy states. It acts as an agent of awareness, linking the unchanging reservoir of the energy pool to the dynamic actions of active energy.

Human Consciousness: Consciousness in humans emerges through the interaction of:

- **The Soul**: Representing dynamic awareness and active energy.

- **The Mind**: Acting as the functional interface for perception and interaction with the physical world.

5. Implications for Understanding the Universe

Bridging Micro and Macro Realms: Energy is shown to operate seamlessly across scales, connecting atomic behavior with cosmic phenomena.

Energy as the Universal Medium: The infinite energy pool binds the cosmos together, shaping and sustaining all phenomena.

Thus, this chapter establishes energy as the ultimate medium underlying all existence, transcending its traditional scientific definitions. By introducing the duality of stable and active energy and highlighting their interactions, this chapter sets the stage for understanding the universe as a unified, dynamic, and interconnected system.

Chapter 3: The Transition

In the infinite energy pool of the universe, transitions between states of energy and matter define the cosmic dance of stability and dynamism. Among the most remarkable manifestations of these transitions are the exotic states of matter that emerge under extreme conditions. **Bose-Einstein Condensates (BEC)** and **Fermionic Condensates** provide a glimpse into the behavior of energy at the edge of its transformation, where the chaotic motions of particles give way to quantum coherence and unity. At near-absolute zero temperatures, these states reveal the underlying principles of the **Unified Energy Dynamics (UED)** framework, showcasing how energy, in its most stable and timeless form, transitions into highly ordered and nearly timeless systems. This chapter explores the profound significance of these condensates as a bridge between the quantum realm and the universal stability of the energy medium.

3.1 Introduction: A Glimpse into Energy's Journey

The universe is a vast, dynamic stage where energy transitions between states, shaping the matter and phenomena that define our reality. At its core, energy exists in two fundamental forms: **stable energy**, timeless and uniform, and **active energy**, dynamic and transformative. The interplay between these forms drives the evolution of the cosmos, from the birth of galaxies to the formation of atomic structures. Among the most intriguing examples of energy transitions are the exotic states of matter that arise under extreme conditions, particularly at temperatures approaching absolute zero.

At this threshold, where motion and chaos nearly cease, energy reveals its capacity for order and coherence, giving rise to states like the **Bose-Einstein Condensate (BEC)** and **Fermionic Condensate**. These states offer a unique perspective on how energy manifests as matter and how universal principles govern even the most extreme conditions.

Energy's Transformation into Matter

Energy transitions into matter through processes governed by stability and dynamism. In the **Unified Energy Dynamics (UED)** framework, matter emerges from localized disturbances within the stable energy medium. These disturbances, driven by active energy, create forms of matter that range from the chaotic (e.g., plasma) to the highly ordered (e.g., solid crystals).

At near-absolute zero temperatures, energy transitions to a state of **minimal activity**:

✓ Particles lose their individual chaotic motion and begin to act in unison.

✓ This transition reflects the influence of the stable energy medium, which supports systems moving toward equilibrium and coherence.

BECs and Fermionic Condensates represent two pathways through which energy organizes itself into matter, each revealing the profound connections between quantum mechanics and universal stability.

Bose-Einstein Condensate (BEC): A Unified Quantum Entity

First predicted by **Albert Einstein** and **Satyendra Nath Bose** in the 1920s, the Bose-Einstein Condensate (BEC) was experimentally realized decades later. This state of matter emerges when **bosons**—particles that follow Bose-Einstein statistics—are cooled to temperatures near absolute zero (−273.15°C).

Key characteristics of a BEC:

- **Quantum Coherence**: At such low temperatures, bosons occupy the same quantum state, behaving as a single quantum entity rather than as individual particles.

- **Minimal Entropy**: The chaotic motion of particles diminishes, and their collective behavior reflects a highly ordered system.

- **Wave Function Overlap**: The wave functions of individual bosons merge, forming a single, unified wave function that spans the entire condensate.

In the UED framework, a BEC represents the transition of active energy into a state closely resembling stable energy. Its quantum coherence aligns with the timeless, harmonious qualities of the stable energy medium.

Fermionic Condensate: Coherence within Constraints

Fermionic Condensates are similar to BECs but involve **fermions**, particles that follow the Pauli exclusion principle (e.g., electrons, protons, and neutrons). Unlike bosons, fermions cannot occupy the same quantum state, making their condensation process more complex. At near-absolute zero, fermions pair up (e.g., Cooper pairs in superconductors) and form a superfluid state where they flow without resistance.

Key characteristics of a Fermionic Condensate:

- **Paired Stability**: Fermions overcome their exclusion principle by forming pairs that behave as composite bosons.

- **Quantum Fluidity**: The system exhibits properties like superfluidity, where resistance to motion vanishes.

- **Intermediate Order**: While distinct from the perfect unification of BECs, fermionic condensates reflect a high degree of coherence and stability.

In UED, Fermionic Condensates highlight the dynamic balance of constraints and coherence, demonstrating how active energy systems can approach the stability of the universal energy medium.

A Window into Universal Dynamics

BECs and Fermionic Condensates provide a rare glimpse into the early universe's potential conditions, where stable energy dominated, and active energy had yet to initiate chaotic interactions. These states reveal:

- **The Pathway to Stability**: As systems cool and chaos diminishes, energy transitions into coherent, nearly timeless states.

- **The Role of the Energy Medium**: The stable energy medium supports the emergence of these states, acting as the framework within which matter organizes itself.

In short, the Bose-Einstein Condensate and Fermionic Condensate exemplify the intricate journey of energy transitioning into matter. At near-absolute zero, these states embody the principles of stability, coherence, and quantum unity, mirroring the timeless properties of the stable energy medium in the UED framework. These exotic states not only deepen our understanding of matter's quantum nature but also provide a gateway to exploring the universe's fundamental energy dynamics.

3.2 The Path to Absolute Zero: Stabilizing Active Energy

The journey to absolute zero represents the pursuit of stability within the vast continuum of energy states. Temperature, as a measure of particle motion and interaction, plays a pivotal role in determining the state of matter. At higher temperatures, active energy dominates, driving the chaotic motion of particles. As temperature decreases, this chaos subsides, and systems move toward equilibrium. Near absolute zero, where thermal energy approaches its minimum, matter reveals properties that closely resemble the **stable energy medium** described in the **Unified Energy Dynamics (UED)** framework.

This section explores the significance of temperature in shaping matter, the unique dynamics near absolute zero, and the theoretical implications for understanding energy transitions within UED.

The Role of Temperature in Determining the State of Matter

Temperature as a Measure of Active Energy: In UED, temperature reflects the degree of **active energy interactions** within a system. Higher temperatures correspond to increased particle motion, energy transitions, and entropy, while lower temperatures indicate reduced activity and greater stability. **For example:** In gases, high temperatures cause particles to move rapidly and collide frequently. In solids, lower temperatures lead to more structured arrangements as particles lose the energy to overcome intermolecular forces.

Phase Transitions and Energy Dynamics: Changes in temperature drive phase transitions, such as the melting of solids, vaporization of liquids, and ionization of gases. These transitions are manifestations of energy reorganization, where active energy interacts with the stable energy medium to produce new states of matter.

Temperature as a Gateway to Quantum Phenomena: As temperature approaches absolute zero ($-273.15^{\circ}C$), quantum effects begin to dominate. Particles exhibit wave-like behaviors, and collective phenomena such as superfluidity and superconductivity emerge.

Approaching Absolute Zero: Minimizing Active Energy Interactions

Theoretical Limit of Absolute Zero: Absolute zero is the point where a system's thermal energy is minimized, and particle motion theoretically ceases. While reaching absolute zero is physically impossible due to quantum fluctuations, achieving near-zero temperatures reveals the underlying stability of matter and energy.

Reduced Chaos and Enhanced Coherence: At near-zero temperatures: Particle motion slows significantly, reducing collisions and chaotic interactions. Systems exhibit **quantum coherence**, where particles behave collectively rather than individually. This state reflects a transition from the dominance of active energy to a condition approaching the stability of the energy medium.

Unique Phenomena Near Absolute Zero:

- **Bose-Einstein Condensates (BECs)**: Particles merge into a single quantum state, behaving as a unified entity.

- **Fermionic Condensates**: Fermions pair up to form superfluid systems, bypassing quantum constraints.

- **Superconductivity and Superfluidity**: Materials exhibit frictionless flow and perfect conductivity, highlighting the near-absence of energy dissipation.

Theoretical Implications for Stable Energy Medium in UED

Stable Energy as the Underlying Framework: The UED framework posits that all active energy interactions occur within a **stable energy medium**. As systems cool and approach absolute zero, the influence of the stable energy medium becomes more apparent: Active energy transitions diminish. Systems converge toward a state of minimal entropy and maximum coherence.

A Glimpse into Timeless Stability: Near absolute zero, matter begins to exhibit properties that align with the timeless stability of the energy medium: Particles lose individuality and behave collectively. Energy dissipation is minimized, resembling the equilibrium state of stable energy. This suggests that the stable energy medium underpins the behavior of matter in these extreme conditions.

Energy Transitions in Reverse: At higher temperatures, active energy dominates, driving chaos and disorder. As temperatures decrease, the system's energy transitions reverse: From chaotic, dynamic states to ordered, coherent systems. From highly active systems to states reflecting stable energy characteristics.

Connection to Universal Energy Dynamics: These near-zero phenomena highlight the interplay between stability and dynamism in the universe: Active energy transitions, which govern most observable phenomena, are minimized. Stability emerges as the dominant force, providing insights into the nature of the infinite energy pool.

Thus, the path to absolute zero is a journey toward stability, where the dominance of active energy interactions gives way to coherence and order. In the **UED framework**, this journey represents a convergence with the **stable energy medium**, offering a glimpse into the timeless foundation of the universe. By studying matter near absolute zero, we can better understand how energy transitions shape the cosmos and how stability emerges as the ultimate state of existence. These insights bridge the quantum behaviors of matter with the universal principles of energy dynamics, deepening our appreciation for the interconnectedness of all phenomena.

3.3 Bose-Einstein Condensate: Unity in Stability

The **Bose-Einstein Condensate** (**BEC**) is a remarkable state of matter that emerges at near-absolute zero temperatures, where the chaotic motion of particles gives way to perfect quantum coherence. In this state, particles known as **bosons** occupy the same quantum state, effectively merging into a single, unified entity. This phenomenon provides profound insights into the nature of energy transitions, stability, and the interplay between quantum mechanics and the **stable energy medium** described in the **Unified Energy Dynamics** (**UED**) framework.

BEC: A State of Perfect Quantum Unity

What is a Bose-Einstein Condensate?

Predicted by **Albert Einstein** and **Satyendra Nath Bose** in the 1920s and realized experimentally in 1995, a BEC forms when a dilute gas of bosons is cooled to temperatures near absolute zero.

At this extreme low temperature, bosons lose their individual identity and occupy the **same quantum state**, behaving as a single quantum entity rather than as independent particles.

Key Characteristics:

- **Quantum Coherence**: The wave functions of the individual particles overlap completely, forming a single, coherent wave function that spans the entire condensate.

- **Minimal Energy State**: The system reaches a state of minimal entropy and maximum stability, with particles exhibiting no relative motion.

- **Superfluidity**: BECs exhibit frictionless flow, a hallmark of quantum coherence and energy conservation.

Behavior of Bosons at Near-Zero Temperatures

What are Bosons?

Bosons are particles with **integer spin** (e.g., photons, helium-4 atoms) that follow **Bose-Einstein statistics.** Unlike fermions, bosons do not obey the Pauli exclusion principle, allowing multiple particles to occupy the same quantum state.

Transition to a BEC: At higher temperatures, bosons behave like classical particles, moving independently and chaotically. As the temperature approaches absolute zero: The thermal energy of the bosons decreases. Their de Broglie wavelengths (quantum wave characteristics) increase and begin to overlap. Eventually, the bosons collapse into a single quantum state, forming a BEC.

Emergent Properties of Bosons in a BEC:

- **Loss of Individuality**: The particles lose their distinct identities and act as a unified system.

- **Wave Function Overlap**: The condensate exhibits a collective behavior dictated by a single quantum wave function.

- **Resistance to Perturbation**: The system's stability makes it resistant to external disturbances, reflecting its near-equilibrium nature.

Connection to the Stable Energy Medium

BECs as a Glimpse into Stability: In the UED framework, the stable energy medium represents a **timeless, equilibrium state**. BECs, with their quantum coherence and minimal entropy, provide a tangible glimpse of this stability. The collective behavior of bosons in a BEC mimics the harmonious characteristics of the stable energy medium.

Energy Transition Toward Stability: The formation of a BEC demonstrates how active energy transitions toward stability: High-temperature systems exhibit chaotic, independent particle motion (active energy). As temperature decreases, interactions become more ordered, leading to the unified stability of a BEC (approaching stable energy).

Timeless Characteristics: In a BEC, the loss of particle individuality and the emergence of a single quantum state create a system that appears almost timeless: The relative motion between particles ceases. The system's behavior is governed by collective, rather than individual, dynamics.

Quantum Unity as a Reflection of Universal Harmony: The coherence in a BEC reflects the **interconnectedness of all energy systems**, a core principle of UED. The transition from active energy to a unified quantum state parallels the cosmic balance between stability and dynamism.

Applications and Implications

Scientific Insights: BECs provide a laboratory for studying quantum mechanics, offering insights into phenomena like superfluidity, superconductivity, and quantum entanglement. They help bridge the gap between microscopic (quantum) and macroscopic (classical) physics.

Technological Potential: BECs hold promise for advancements in quantum computing, precision measurement, and the development of ultra-sensitive detectors.

Cosmic Relevance: BECs may model the conditions of the early universe, where energy was stable and undifferentiated before transitioning into active forms of matter and radiation.

Thus, the **Bose-Einstein Condensate** is a profound demonstration of energy's ability to achieve unity and stability under extreme conditions. By aligning with the principles of the **stable energy medium** in the UED framework, BECs bridge the quantum and universal scales, offering a glimpse into the harmonious interplay between energy states. Their existence highlights how energy, when sufficiently cooled and ordered, transitions into a timeless, unified state—a reflection of the ultimate stability at the heart of the universe.

3.4 Fermionic Condensate: Coherence within Constraints

The **Fermionic Condensate** is a fascinating state of matter that emerges under conditions of extreme cold, where fermions—particles that inherently resist sharing the same quantum state—achieve quantum coherence through pairing mechanisms. Unlike **Bose-Einstein Condensates (BECs)**, where bosons unify into a single quantum state, fermionic condensates reflect a more intricate process dictated by the **Pauli exclusion principle**. In the context of **Unified Energy Dynamics (UED)**, fermionic condensates embody the balance between stability and dynamism, showcasing how ordered energy dynamics arise even within constrained systems.

Fermions and the Pauli Exclusion Principle

What are Fermions?

Fermions are fundamental particles that make up matter, including electrons, protons, and neutrons.

They are characterized by having **half-integer spins** (e.g., 1/2,3/2) and follow **Fermi-Dirac statistics.**

The Pauli Exclusion Principle: Proposed by Wolfgang Pauli in 1925, this principle states that no two fermions can occupy the same quantum state simultaneously. This principle defines the structure of atoms and matter, as electrons occupy distinct energy levels around a nucleus.

Challenges of Condensation for Fermions: Unlike bosons, which readily occupy the same quantum state, fermions require creative pathways to achieve condensation due to their inherent quantum constraints.

Formation of Cooper Pairs and Quantum Coherence

The Role of Pairing in Overcoming Constraints: At near-absolute zero temperatures, fermions pair up to form **Cooper pairs**, which behave like composite bosons. This pairing is critical for achieving condensation and quantum coherence.

What are Cooper Pairs?

First described in the **BCS theory** (Bardeen-Cooper-Schrieffer theory) of superconductivity, Cooper pairs are formed when two fermions (e.g., electrons) interact indirectly via lattice vibrations (phonons) in a material.

Despite the fermions' individual repulsion due to their like charges, the lattice vibrations create an attractive force, enabling them to pair up.

Quantum Coherence in Fermionic Systems: Once paired, fermions behave collectively, exhibiting quantum coherence similar to that seen in BECs. The system transitions into a **superfluid** or **superconducting** state, where: Resistance to motion vanishes (e.g., electrons flow without resistance in superconductors). Energy dissipation is minimized, reflecting a state of ordered energy dynamics.

Stability in Fermionic Condensates as a Reflection of Ordered Energy Dynamics

Order amidst Constraints: Fermionic condensates demonstrate how systems with inherent restrictions can still achieve stability and coherence through collective behavior. In the UED framework, this reflects the principle that even active energy systems constrained by quantum mechanics can align with the underlying stable energy medium.

Energy Minimization: The formation of Cooper pairs reduces the system's energy to a near-minimal state, mirroring the low-entropy characteristics of the stable energy medium. The absence of friction or resistance in these systems reflects their alignment with the principles of universal stability.

Transition to Stability: The progression from chaotic, high-energy particle interactions to the ordered state of a fermionic condensate exemplifies energy's journey toward equilibrium, a core tenet of UED.

Implications for Science and UED

Insights into Quantum Mechanics: Fermionic condensates provide a deeper understanding of quantum phenomena, such as superconductivity and superfluidity, where energy transitions result in macroscopic quantum effects.

Applications in Technology: Technologies like **quantum computers**, **MRI machines**, and advanced energy systems benefit from the principles of fermionic condensation and quantum coherence.

Cosmic and Universal Relevance: Fermionic condensates may model the behavior of matter in extreme cosmic environments, such as the cores of neutron stars, where fermions are compressed into degenerate states. These condensates bridge the gap between localized quantum phenomena and the universal dynamics of energy, aligning with UED's holistic vision.

Connection to the UED Framework

Energy Interplay in Fermionic Systems: The pairing of fermions into Cooper pairs reflects a dynamic interplay between active energy (individual particle motion) and stability (collective coherence). In UED, this represents how the stable energy medium influences even highly constrained systems to achieve order.

Reflection of Stable Energy Principles: Fermionic condensates approach the characteristics of the stable energy medium by minimizing entropy and aligning particle behavior, revealing universal principles at play in quantum systems.

Timeless Characteristics in Quantum Coherence: The timeless behavior of fermionic condensates, where energy dissipation ceases, provides a glimpse into the equilibrium state of stable energy.

Thus, Fermionic condensates exemplify the elegance of ordered energy dynamics within the constraints of quantum mechanics. By forming Cooper pairs and achieving quantum coherence, fermions demonstrate how systems with inherent limitations can transition toward stability. In the **Unified Energy Dynamics (UED)** framework, fermionic condensates reflect the influence of the stable energy medium on active energy systems, bridging the microcosmic world of quantum phenomena with the macrocosmic principles of universal harmony. These states of matter not only deepen our understanding of energy transitions but also highlight the interconnectedness of all forms of energy in the cosmos.

3.5 The Role of Stable Energy in Exotic States

The **Bose-Einstein Condensate (BEC)** and **Fermionic Condensate** are profound examples of how matter transitions into highly stable and coherent states under extreme conditions. These exotic states of matter highlight the underlying influence of **stable energy**, as described in the **Unified Energy Dynamics (UED)** framework. Stable energy, the timeless and uniform substrate of the universe, governs the behavior of systems as they approach equilibrium, enabling the emergence of coherence and order from chaos. This section explores how stable energy underpins the formation of these condensates, the role of entropy reduction in achieving quantum coherence, and the insights these states offer into the universal principles of stability.

How Stable Energy Underpins the Formation of BEC and Fermionic Condensate

Stable Energy as the Universal Foundation: In UED, stable energy represents the equilibrium state of the energy medium, free from disturbances and transitions. Both BEC and Fermionic Condensates form as active energy within a system diminishes, aligning the system more closely with the stable energy medium.

Energy Transitions Toward Stability: As systems are cooled toward absolute zero: The thermal energy driving chaotic particle motion diminishes. Particles begin to behave collectively, reflecting the coherence and stability of the energy medium. This transition is a journey from active energy (dynamic, disordered) to a state approaching stable energy (timeless, ordered).

Mechanisms in BEC and Fermionic Condensates: In **BECs**, bosons condense into a single quantum state, behaving as a unified whole. This reflects the stable energy medium's timeless and collective properties. In **Fermionic Condensates**, fermions overcome quantum constraints (Pauli exclusion principle) by forming Cooper pairs, achieving coherence through collective interaction.

The Reduction of Entropy and Emergence of Quantum Coherence

Entropy as a Measure of Disorder: In high-energy systems, entropy is a reflection of the chaotic and independent behavior of particles. As temperature decreases and systems approach absolute zero, entropy reduces significantly, signaling a shift toward order and coherence.

BEC and Fermionic Condensates as Low-Entropy States:

- **BECs**: The wave functions of bosons overlap completely, eliminating distinctions between particles and minimizing entropy.

- **Fermionic Condensates**: The pairing of fermions reduces energy dissipation, creating a frictionless, low-entropy state.

Quantum Coherence and Stability: Quantum coherence emerges as particles act collectively, forming a system governed by a single wave function. This coherence is a hallmark of systems influenced by the stable energy medium, where interactions transition from individual to collective behavior.

Alignment with the Stable Energy Medium: The reduction of entropy and the emergence of coherence reflect the system's alignment with the stable energy medium's equilibrium state. These exotic states offer a glimpse into the timeless and ordered nature of stable energy.

Insights into Universal Stability from These States

The Path to Equilibrium: BECs and Fermionic Condensates demonstrate how systems, even under constraints, move toward equilibrium when active energy diminishes. This mirrors the universal process of energy seeking balance, a fundamental principle of UED.

Stability Amid Complexity: In **BECs**, stability arises from the unification of bosons into a single quantum state, eliminating individuality. In **Fermionic Condensates**, stability emerges through the pairing of fermions, overcoming inherent quantum constraints. These states illustrate how stability can emerge through different pathways, even in systems governed by distinct rules.

Universal Implications: The behavior of these exotic states offers clues about the early universe, where stable energy dominated before the emergence of active energy and matter. They reflect the principles of universal harmony and interconnectedness, where localized energy systems align with the stable energy medium.

Bridge Between Micro and Macro Scales: The coherence observed in these condensates parallels the coherence seen in large-scale cosmic structures, such as galaxies and black holes. This suggests a universal principle of stability that operates across scales, linking quantum phenomena with cosmic dynamics.

Thus, the **Bose-Einstein Condensate (BEC)** and **Fermionic Condensate** reveal the profound role of **stable energy** in shaping matter under extreme conditions. These states, characterized by low entropy and quantum coherence, align closely with the timeless stability of the energy medium described in UED. By studying these condensates, we gain insights into the universal principles of equilibrium, coherence, and interconnectedness, bridging the quantum realm with the vast dynamics of the cosmos. These exotic states serve as tangible manifestations of UED's foundational concepts, illustrating how energy transitions toward stability in the infinite energy pool of the universe.

3.6 Implications for the Early Universe

The **Bose-Einstein Condensate (BEC)** and **Fermionic Condensate** offer profound insights into the conditions and dynamics of the early universe. These exotic states, formed under near-absolute zero conditions, provide a unique window into how energy transitions into matter, coherence, and structure. By drawing parallels between the properties of these condensates and the universe's formative stages, we can explore how stability and dynamism shaped the cosmic evolution described in the **Unified Energy Dynamics (UED)** framework.

Parallels Between Early Universe Conditions and Exotic States

The Early Universe as an Energy-Dominant State: In its earliest moments, the universe existed as an energy-dominated system, where matter had not yet emerged. Energy existed in its most stable form, with minimal active transitions, reflecting conditions analogous to the near-equilibrium states seen in BECs and Fermionic Condensates.

Extreme Conditions at the Cosmic Beginning: Similar to the extreme cold required to form a BEC or Fermionic Condensate, the early universe's conditions were extreme but in the opposite direction—high energy density and temperature. As the universe expanded and cooled, localized transitions began, leading to the formation of particles, atoms, and eventually large-scale structures.

Timelessness and Coherence: Both the early universe and exotic states exhibit characteristics of timelessness and coherence: In the early universe, quantum fluctuations governed initial dynamics, with time emerging as energy transitions began. In BECs and Fermionic Condensates, the loss of individual particle identities and the emergence of quantum coherence mirror the universe's initial, unified energy state.

Energy Transitions During Cosmic Evolution

From Stability to Dynamism: The UED framework describes the universe as transitioning from a state of **stable energy** to **active energy**, driving the formation of matter and observable phenomena. Exotic states like BECs and Fermionic Condensates model this transition: BECs reflect the unified, timeless state of stable energy. Fermionic Condensates demonstrate how constraints (like the Pauli exclusion principle) lead to structured dynamism, akin to the emergence of particles and atoms.

Formation of Quantum Structures: In the early universe, energy transitions led to the formation of quarks, gluons, and other fundamental particles, eventually creating protons, neutrons, and electrons. BECs and Fermionic Condensates replicate similar processes on a smaller scale, where particles transition from chaotic motion to coherent states, forming quantum structures governed by universal principles.

Entropy and Coherence: The early universe's expansion increased entropy, allowing energy to diversify into active forms. However, local regions retained coherence, forming stars, galaxies, and black holes. Exotic states demonstrate the inverse process, where entropy is minimized, and coherence emerges as systems approach equilibrium.

Reflection of Universal Principles in Exotic States

Unified Energy Dynamics: Exotic states provide evidence for UED's principles of energy interplay:

- **BECs**: Reflect the stable energy medium's timeless properties, where active transitions are nearly absent.

- **Fermionic Condensates**: Showcase the balance between stability and active constraints, mirroring the universe's structured complexity.

Microscopic Insights into Macroscopic Phenomena: The coherence in BECs and Fermionic Condensates parallels large-scale phenomena, such as: The alignment of cosmic microwave background (CMB) radiation. The organized motion of galaxies and cosmic filaments. These parallels suggest that the same principles govern the universe across all scales.

Timelessness and Time Emergence: In exotic states, the loss of particle individuality and motion creates a nearly timeless system. Similarly, in the early universe, time emerged as energy transitions began, marking the shift from a timeless to a dynamic state.

How BECs and Fermionic Condensates Illuminate Cosmic Evolution

Modeling the Early Universe: BECs and Fermionic Condensates act as laboratories for studying the early universe's energy dynamics: They replicate conditions where energy transitions are minimal or highly constrained. Their quantum coherence and stability provide insights into how energy coalesced into matter.

Understanding Structure Formation: The transitions in exotic states—from chaotic to ordered—mirror the processes that led to the formation of stars, galaxies, and other cosmic structures.

Clues About Dark Matter and Energy: The behavior of fermions in condensates offers analogies for understanding dark matter, which interacts weakly with normal matter but contributes significantly to cosmic structure. The timeless, stable properties of BECs might parallel aspects of dark energy, the mysterious force driving the universe's accelerated expansion.

In summary, exotic states of matter, such as **Bose-Einstein Condensates (BECs)** and **Fermionic Condensates**, offer profound insights into the conditions and dynamics of the early universe. By reflecting the transition from stable to active energy, these states provide a model for understanding cosmic evolution, from the formation of fundamental particles to the emergence of large-scale structures. In the UED framework, these states exemplify the universal principles of stability, coherence, and energy interplay, bridging the microcosmic behaviors of quantum systems with the macrocosmic dynamics of the cosmos. Through these condensates, we gain a deeper appreciation for the timeless forces that shape our universe.

3.7 Applications in Science and Technology

The discovery and exploration of **Bose-Einstein Condensates (BECs)** and **Fermionic Condensates** have opened new frontiers in science and technology, providing groundbreaking insights into quantum mechanics and energy dynamics. These states not only deepen our understanding of the universe but also have significant implications for next-generation technologies, such as quantum computing and superfluidity. Within the **Unified Energy Dynamics (UED)** framework, BECs and Fermionic Condensates serve as practical demonstrations of UED principles, bridging theoretical concepts with real-world applications.

Quantum Computing: Harnessing Coherence for Processing Power

Quantum Coherence in Condensates: Both BECs and Fermionic Condensates exhibit **quantum coherence**, where particles behave collectively as a single quantum state. This coherence is essential for quantum computing, where quantum bits (qubits) must maintain stable superpositions to perform computations.

Enhanced Qubit Stability: The principles of BECs and Fermionic Condensates can improve the stability and coherence times of qubits, reducing errors in quantum computations. Techniques inspired by these states, such as pairing particles or using bosonic systems, could lead to more robust quantum processors.

Scalable Quantum Systems: BECs, with their unified wave functions, provide a model for creating scalable and highly coherent quantum systems, essential for advancing quantum computing.

Superfluidity: Frictionless Flow for Advanced Materials

What is Superfluidity? Superfluidity, observed in BECs and Fermionic Condensates, is the property of a fluid to flow without viscosity or energy loss. This phenomenon emerges from quantum coherence, where particles move collectively without individual resistance.

Applications in Material Science: Superfluid systems could revolutionize the development of advanced materials with near-zero energy dissipation. These materials could be used in ultra-efficient energy transport systems, such as superconductors or frictionless bearings.

Cryogenics and Cooling: Superfluidity has applications in cryogenics, where it enables efficient cooling mechanisms for quantum systems and advanced sensors.

Next-Generation Technologies

Precision Measurement Devices: The sensitivity of BECs and Fermionic Condensates to external forces makes them ideal for developing **precision measurement devices**: Atomic clocks with unprecedented accuracy. Gravitational wave detectors that rely on quantum coherence for enhanced sensitivity.

Sensors for Environmental Monitoring: Condensates can detect minute changes in magnetic, electric, or gravitational fields, paving the way for ultra-sensitive environmental sensors.

Transport and Energy Systems: Superfluid and superconducting systems could lead to: Lossless power grids, significantly reducing energy waste. Advanced propulsion systems for spacecraft using low-energy, high-efficiency mechanisms.

Biomedical Applications

Imaging and Diagnostics: BECs and related technologies contribute to the development of high-resolution imaging systems, such as quantum-enhanced MRI scanners. These technologies improve diagnostic capabilities, particularly in detecting subtle anomalies in biological tissues.

Drug Delivery Systems: Understanding the quantum properties of condensates could lead to innovations in targeted drug delivery, where energy transitions are harnessed for precision.

Demonstrating UED Principles in Practical Terms

Stability and Coherence in Energy Systems: The formation of BECs and Fermionic Condensates showcases how **stable energy principles** can be applied to achieve order and coherence in dynamic systems. These states demonstrate the transition from chaotic energy systems to stable, predictable behaviors, aligning with UED's core tenets.

Energy Efficiency and Minimal Entropy: The low-entropy characteristics of these states inspire the design of systems that minimize energy dissipation and maximize efficiency. Technologies such as superconductors and quantum computers are direct applications of these principles.

Bridging Quantum and Cosmic Scales: By modeling universal stability at the quantum level, BECs and Fermionic Condensates provide a tangible link to the UED framework's description of energy dynamics at cosmic scales.

Future Directions in Science and Technology

Quantum Networks: Leveraging the coherence of BECs, future quantum networks could achieve near-instantaneous communication over vast distances, revolutionizing information transfer.

Space Exploration: Superfluidity and zero-resistance systems inspired by these states could lead to more efficient propulsion and energy storage for interstellar missions.

Sustainable Energy Solutions: The principles of low entropy and high coherence could guide the development of sustainable energy systems, reducing waste and enhancing output.

In short, the practical applications of **Bose-Einstein Condensates** and **Fermionic Condensates** demonstrate the profound connection between theoretical physics and cutting-edge technology. From quantum computing and superfluidity to next-generation materials and biomedical advances, these states embody the principles of stability, coherence, and energy interplay central to the **Unified Energy Dynamics (UED)** framework. As we continue to explore and harness these exotic states, they promise to revolutionize technology, bridge quantum and cosmic understanding, and create a future shaped by the harmonious dynamics of energy.

3.8 Conclusion: The Timeless Connection

The **Bose-Einstein Condensate (BEC)** and **Fermionic Condensate** are profound manifestations of energy's journey, exemplifying the balance between stability and dynamism at the heart of the **Unified Energy Dynamics (UED)** framework. These exotic states, emerging at near-absolute zero temperatures, provide a glimpse into the fundamental principles governing the universe, where energy transitions into highly ordered and coherent systems. By bridging the quantum and cosmic scales, they illuminate the intricate interplay between stable and active energy, revealing the timeless connection that binds the universe together.

Key Examples of UED Principles

Energy Interplay in Transition States: Both BECs and Fermionic Condensates arise from the cooling of active energy systems, reducing thermal chaos and driving particles toward coherence and order. In BECs, bosons merge into a single quantum state, reflecting the unifying nature of stable energy. In Fermionic Condensates, fermions achieve coherence through pairing mechanisms, overcoming quantum constraints to align with stability.

Stable Energy Medium as the Foundation: These states embody the characteristics of the **stable energy medium**:

✓ **Timelessness**: In their near-absolute zero state, particles exhibit minimal motion, creating a system that appears almost timeless.

✓ **Coherence**: Quantum coherence in these systems mirrors the stability and order of the universal energy pool.

Entropy Reduction and Universal Order: The dramatic reduction in entropy in these states highlights energy's natural tendency to seek equilibrium. This aligns with UED's core principle that all energy systems strive toward balance and stability, even under extreme constraints.

Bridging Scales: BECs and Fermionic Condensates demonstrate how universal principles operate across scales: At the quantum level, they reveal how energy transitions from chaotic particle motion to ordered systems. At the cosmic scale, they reflect the coherence and stability seen in phenomena like galactic alignment or black hole dynamics.

Significance in Understanding Energy's Journey

Pathways to Stability: These states exemplify energy's journey from dynamic, high-entropy systems to ordered, low-entropy configurations. They provide a model for understanding how energy transitions between stable and active forms, offering insights into both quantum phenomena and cosmic evolution.

Insights into the Early Universe: BECs and Fermionic Condensates parallel the conditions of the early universe, where energy existed in a stable, undifferentiated state before transitioning into matter and motion. By studying these states, we gain a deeper understanding of the processes that shaped the cosmos from its initial moments.

Practical and Technological Relevance: The principles demonstrated by these states are not merely theoretical—they underpin advancements in quantum computing, superfluidity, and sustainable energy technologies. They serve as a bridge between theoretical physics and real-world applications, showcasing the practical implications of UED principles.

Cosmic Harmony: These states embody the timeless connection between stability and activity, highlighting the universe's inherent harmony. They reflect UED's vision of the cosmos as a dynamic system governed by a balance of stability and transformation.

The Timeless Connection

Energy's Fundamental Duality: BECs and Fermionic Condensates reaffirm the dual nature of energy in UED: Stable energy provides the foundation and equilibrium. Active energy drives transitions and observable phenomena.

Unified Framework for Understanding: These states integrate the microcosmic and macrocosmic scales, offering a unified perspective on energy dynamics. They illustrate how the same principles govern the behavior of particles and the evolution of galaxies.

A Glimpse into Universal Stability: By reducing entropy and achieving coherence, these states provide a glimpse into the universal stability that underpins all existence. They remind us that the journey of energy—though dynamic and transformative—always seeks balance within the infinite energy medium.

The **Bose-Einstein Condensate** and **Fermionic Condensate** are not just exotic states of matter—they are profound demonstrations of the principles at the core of the **Unified Energy Dynamics (UED)** framework. By transitioning from chaos to coherence, they reveal the timeless connection between stability and activity that governs the universe. These states bridge quantum mechanics and cosmic phenomena, offering insights into energy's journey and the universal harmony that sustains all existence. Their significance extends beyond theoretical physics, influencing technology, science, and our understanding of the cosmos itself. Through these condensates, we glimpse the infinite potential of energy, guided by the principles of stability, coherence, and timeless balance.

Chapter 4: The Nature of Forces

As we continue our exploration of energy as the universal medium, we turn to one of its most tangible manifestations: forces. Forces govern the interactions that shape our reality, from the binding of atoms to the motion of galaxies. Traditionally understood as distinct phenomena—gravitational, electromagnetic, and nuclear—forces have been studied as the fundamental drivers of change in both the microscopic and macroscopic realms. Yet, when viewed through the lens of Unified Energy Dynamics, forces emerge not as isolated entities but as expressions of energy transitions, deeply interconnected and unified by the principles of stability and dynamism.

This chapter delves into the essence of forces, redefining their nature as manifestations of energy in action. From gravity's interaction with the stable energy medium to the cohesive thread uniting all forces, we explore how these phenomena arise, evolve, and sustain the harmony of the cosmos. Through this unified perspective, we seek to unravel the illusion of separation, revealing forces as integral components of a single, universal energy framework.

4.1 Introduction: Forces as Expressions of Energy

Forces are among the most fundamental aspects of the universe, shaping the interactions of matter and the behavior of systems across all scales. From the gravitational pull that binds planets to stars, to the electromagnetic forces that govern the intricate structures of atoms, forces are traditionally viewed as distinct entities, each with its unique characteristics and governing equations. However, when seen through the lens of **Unified Energy Dynamics**, forces transcend their conventional definitions and emerge as expressions of energy transitions—a manifestation of the dynamic interplay between stability and motion.

Forces as Emergent Phenomena of Energy Dynamics

At their core, forces are not independent entities but rather emergent phenomena arising from energy's dual nature. They reflect the interactions and transitions between **stable energy** (the reservoir that provides coherence and structure) and **active energy** (the dynamic force driving motion and transformation). Forces can be understood as the "language" through which energy communicates and influences the behavior of matter and systems in the universe.

Active Energy in Motion: Forces like electromagnetism and nuclear interactions arise from active energy's dynamic actions. They drive observable phenomena such as the binding of electrons to nuclei, the fusion of stars, and even the transmission of information via light waves.

Stable Energy as the Framework: Forces depend on the stability provided by the universal energy pool. Gravity, for instance, reflects not just the attraction between masses but also the interaction of mass with the stable energy medium, creating the architecture of spacetime itself.

This perspective shifts the understanding of forces from isolated influences to interconnected expressions of a unified energy medium. By studying forces through this lens, we uncover their deeper role in maintaining the balance and harmony of the cosmos.

Bridging Traditional and Unified Perspectives on Forces

The traditional understanding of forces has been remarkably successful in explaining natural phenomena within specific domains:

Newtonian Mechanics: Gravity is treated as a mass-based attraction, elegantly described by Newton's law of universal gravitation, which predicts the motion of celestial bodies with precision.

Electromagnetism: Maxwell's equations reveal the interplay between electric and magnetic fields, explaining phenomena ranging from the behavior of light to the operation of electric circuits.

Quantum Mechanics: At subatomic scales, forces like the strong and weak nuclear interactions describe particle interactions, ensuring the stability of matter and driving processes like radioactive decay.

Einstein's Relativity: Gravity is reimagined as the curvature of spacetime caused by mass and energy, offering profound insights into phenomena like black holes and the expansion of the universe.

While these frameworks provide extraordinary insights, they are limited by fragmentation, with each force described separately and often without connection to others.

Unified Energy Dynamics bridges this gap by revealing the underlying principles that connect all forces:

✓ **Forces as Transitions**: Forces are the mechanisms by which energy transitions occur, whether through motion, interaction, or transformation.

✓ **Universal Medium**: All forces operate within the infinite energy pool, influenced by the same fundamental principles of stability and dynamism.

✓ **Interconnection and Unity**: The apparent separation of forces is an illusion born of context and scale. At their essence, all forces are unified expressions of the same energy dynamics, shaped by local and universal conditions.

Toward a Unified Understanding of Forces

By redefining forces as emergent phenomena of energy, we can connect the macroscopic with the microscopic, the classical with the quantum, and the physical with the metaphysical. This unified approach not only reconciles existing theories but also opens new pathways for understanding and exploration.

In the chapters to follow, we will:

✓ Examine the nature of gravity as an interaction not just between masses but with the stable energy medium itself.

✓ Explore how the seemingly distinct forces of nature—gravitational, electromagnetic, and nuclear—emerge as facets of the same energy framework.

✓ Address the illusion of separation, showing that all forces are interconnected expressions of the universe's infinite energy pool.

By doing so, we aim to present a cohesive view of forces as the dynamic expressions of energy transitions, integral to the universal harmony that binds all existence.

4.2 Forces as Manifestations of Energy Transitions

Forces, traditionally considered distinct entities governing the interactions of matter, are better understood as manifestations of energy transitions. These transitions occur within the infinite energy pool, arising from the interplay of **stable energy** (the universal reservoir of latent potential) and **active energy** (the dynamic driver of motion and transformation). Forces are the observable effects of this interaction, shaping reality from the quantum scale to the vastness of the cosmos.

Energy Interactions as the Root of All Forces

At the most fundamental level, forces are not isolated phenomena but expressions of how energy transitions shape interactions within and between systems. These interactions are governed by the constant exchange, transformation, and redistribution of energy:

Stable Energy as a Foundation: Stable energy provides the structural coherence necessary for the emergence of forces. It acts as the medium through which forces propagate, ensuring the continuity and consistency of interactions across space and time.

Active Energy as a Catalyst: Active energy drives change by initiating movement, interaction, and transformation. It manifests in forms such as electromagnetic radiation, thermal energy, and kinetic motion, creating the effects we observe as forces.

Forces as Expressions of Transition: Forces emerge when active energy interacts with the stable energy reservoir or between localized pockets of active energy. For instance, gravitational forces arise from the interaction of mass with the stable energy medium, while electromagnetic forces reflect the exchange of energy between charged particles.

These interactions highlight the interconnectedness of forces and their reliance on the fundamental principles of energy dynamics.

How Stable and Active Energy Interplay Gives Rise to Forces

The dual nature of energy—its simultaneous states of stability and dynamism—provides the framework for understanding how forces arise. This interplay creates the conditions necessary for forces to manifest, guiding the behavior of matter and energy across scales:

Gravity as an Interaction with Stability: Gravity, often viewed as an attractive force between masses, is better understood as an interaction between mass and the stable energy medium. The presence of mass disturbs the equilibrium of the stable energy pool, creating a gradient that manifests as gravitational effects. This interaction is not merely an attraction but a response of the stable medium to the presence of localized energy.

Electromagnetic Forces as Dynamic Energy Interplay: Electromagnetic forces arise from the motion and interaction of charged particles, driven by active energy. These forces govern atomic and molecular structures, enabling the formation of matter. The stable energy medium ensures the propagation of electromagnetic waves, maintaining coherence as these waves transfer energy across vast distances.

Nuclear Forces as Localized Transitions: The strong and weak nuclear forces operate at the quantum scale, arising from intense energy interactions within atomic nuclei. The strong force binds protons and neutrons, while the weak force drives processes like radioactive decay. These forces exemplify the interplay of stability and motion at the smallest scales, ensuring the stability of matter while enabling transformation.

Observable Effects Across Scales: From Quantum to Cosmic

The manifestations of forces, driven by energy transitions, can be observed across all scales of existence. This universality underscores the fundamental role of energy dynamics in shaping the cosmos:

At the Quantum Scale: Forces manifest as interactions between particles, mediated by energy carriers like photons and gluons. These interactions govern phenomena such as quantum entanglement and the exchange of energy in subatomic processes. The dual nature of energy is evident in the wave-particle duality of photons and electrons, where stable energy provides coherence and active energy drives motion.

At the Human Scale: Forces like gravity, friction, and electromagnetism shape everyday experiences, from the motion of objects to the transmission of light and sound. These observable effects are direct outcomes of energy transitions within localized systems, governed by the interplay of stability and motion.

At the Cosmic Scale: Gravitational forces create the architecture of the universe, shaping galaxies, stars, and planetary systems. These effects arise from the interaction of mass and the stable energy medium, creating curvature in spacetime. Electromagnetic forces drive cosmic phenomena such as the radiation of light from stars and the propagation of signals across interstellar distances.

The Universal Nature of Forces

Understanding forces as manifestations of energy transitions unifies their diverse effects, revealing a consistent framework that spans all scales of existence. This perspective dissolves the apparent distinctions between forces, highlighting their shared origin in the interplay of stable and active energy.

Through this lens, forces emerge as the language of energy—a dynamic expression of its transitions that governs the interactions, transformations, and coherence of the universe. By examining forces in this unified context, we gain deeper insights into the nature of reality, uncovering the principles that bind the cosmos into a single, interconnected system.

4.3 Redefining Gravity: Beyond Mass-Based Attraction

Gravity, the most familiar yet enigmatic force in the universe, has traditionally been understood as the attractive force between masses. This classical view, pioneered by Newton and expanded upon by Einstein's General Relativity, explains gravity in terms of mass, distance, and spacetime curvature. However, through the lens of Unified Energy Dynamics (UED), gravity emerges not merely as an attraction between masses but as a manifestation of energy interactions within the infinite, stable energy medium. This perspective shifts our understanding of gravity from a force rooted solely in mass to a phenomenon governed by the dynamic interplay of energy and stability.

Gravity as an Interaction between Mass and the Stable Energy Medium

In the framework of Unified Energy Dynamics, gravity is fundamentally redefined as the interaction between mass and the stable energy medium that permeates the universe. This reinterpretation resolves several conceptual challenges associated with classical and relativistic theories of gravity:

Mass Disturbs the Energy Medium: Mass, as a localized concentration of active energy, creates a disturbance in the stable energy pool. This disturbance generates gradients within the energy medium, which we perceive as gravitational effects. The presence of mass bends the stable energy medium, much like a heavy object placed on a taut sheet creates depressions. This bending is the origin of gravitational influence, with objects moving along the paths defined by these energy gradients.

Gravity as a Response to Energy Disequilibrium: Gravity is not an inherent property of mass but a response of the stable energy medium to the presence of mass. It seeks to restore equilibrium within the energy pool. The gravitational "force" arises as the stable energy medium adjusts to the localized energy imbalance created by mass, pulling objects together to stabilize the disturbance.

Gravitational Waves and Energy Interactions: Phenomena like gravitational waves, ripples in spacetime caused by massive accelerating objects, reflect the interaction between active energy (mass in motion) and the stable energy pool. These waves demonstrate how gravity is not merely an attractive force but a dynamic interplay of energy transitions within the infinite energy medium.

The Role of Energy Dynamics in Shaping Gravitational Behavior

Gravity's behavior is shaped by the intricate dynamics of stable and active energy. By focusing on energy transitions, UED provides a cohesive explanation of gravitational phenomena across scales:

Localized Effects: At small scales, such as planetary systems, gravity appears as a predictable force proportional to mass and inversely proportional to the square of the distance. This observation aligns with Newtonian mechanics but is explained here as the localized adjustment of the stable energy medium to concentrated active energy (mass).

Universal Effects: On a cosmic scale, gravity shapes the structure of galaxies, binds stars into clusters, and governs the motion of celestial bodies. These effects arise from large-scale energy interactions where the stable energy medium maintains the coherence of the universe's architecture.

Weakest Force on Universal Scale: Gravity's relative weakness compared to other forces, such as electromagnetism and the nuclear forces, becomes evident when viewed as an energy interaction. While it is the dominant force at macroscopic scales due to the additive nature of mass, its dependence on the stable energy medium explains its subtlety at quantum levels.

Interaction with Light and Energy: Gravitational lensing, where light bends around massive objects, illustrates gravity's effect on active energy forms (like photons) as they traverse the gradients in the stable energy medium.

Revisiting Classical Gravity through the Lens of Unified Energy Dynamics

The classical understanding of gravity, while effective for practical applications, encounters limitations when faced with extreme conditions or at quantum scales. By integrating UED, these challenges are reframed and addressed:

Moving Beyond Newton's Attraction: Newton's laws describe gravity as an attractive force between two masses, but they do not explain the mechanism behind this attraction. UED identifies this mechanism as the disturbance mass creates in the stable energy medium, offering a deeper, energy-centric understanding.

Expanding Einstein's Curvature: Einstein's General Relativity explains gravity as the curvature of spacetime caused by mass and energy. UED complements this view by focusing on the interaction between the stable energy medium and active energy forms, emphasizing the role of energy transitions in shaping spacetime curvature.

Implications for Black Holes: In classical and relativistic physics, black holes represent regions of intense gravity where spacetime curvature becomes infinite. In UED, black holes are understood as points where active energy dissolves into the stable energy pool, highlighting the role of energy dynamics in these extreme environments.

Rethinking Gravitational Anomalies: Observations of phenomena like dark matter and dark energy suggest gravitational behavior that deviates from classical predictions. UED provides a new lens for interpreting these anomalies, suggesting they may arise from interactions within the stable energy medium rather than unobserved mass.

In summary, Gravity, when redefined as an interaction between mass and the stable energy medium, becomes more than an attractive force—it becomes a manifestation of the universe's intrinsic energy dynamics. This perspective unifies classical and relativistic views, grounding gravitational phenomena in the interplay of stable and active energy. By understanding gravity in this way, we unlock a deeper appreciation of its role in shaping the cosmos, from the motion of planets to the structure of galaxies and beyond. This reinterpretation also paves the way for integrating gravity into a broader unified framework, linking it with other forces and phenomena within the infinite energy pool.

4.4 Electromagnetic and Nuclear Forces: Dynamic Energy Interactions

The forces of nature, often perceived as distinct, are unified in their origin within the interplay of energy dynamics. While gravity governs large-scale cosmic interactions, the electromagnetic and nuclear forces dominate the microscopic realm, driving interactions at the atomic and subatomic levels. These forces, when viewed through the lens of Unified Energy Dynamics (UED), emerge as manifestations of energy transitions, seamlessly connecting the quantum scale with the universal energy pool.

The Electromagnetic Force: Active Energy in Motion

Electromagnetic force, one of the fundamental forces, is the embodiment of active energy in motion. Its influence spans vast scales, from governing atomic structures to enabling macroscopic phenomena such as light and electricity.

Energy and Charge Interaction: Electromagnetic force arises from the interaction of charged particles, such as electrons and protons, with the surrounding stable energy medium. It manifests as attraction or repulsion between particles based on their charge, creating dynamic energy fields that influence their motion and behavior.

Wave-Particle Duality in Action: Light and other electromagnetic waves exemplify active energy in motion. These waves propagate through the stable energy medium, demonstrating energy's capacity to transition between states while interacting with matter. The dual nature of electromagnetic waves, acting as both particles (photons) and waves, reflects the interplay between active energy's dynamic presence and its interaction with stable energy.

Universal Connectivity: Electromagnetic waves serve as carriers of energy and information, bridging the micro and macro realms. From powering biological processes in photosynthesis to enabling cosmic communication through radio waves, this force exemplifies energy's role in connecting disparate systems. The propagation of light through the vastness of space underscores the omnipresence of the stable energy medium, enabling the motion of active energy across unimaginable distances.

Strong and Weak Nuclear Forces: Stabilizers at the Quantum Scale

At the heart of matter, the strong and weak nuclear forces operate as the stabilizing agents of energy within the atomic nucleus, ensuring coherence amidst the immense energies involved.

The Strong Nuclear Force: The strong force binds protons and neutrons within the nucleus, overcoming the electromagnetic repulsion between positively charged protons. This force operates at minuscule distances, ensuring stability within the atomic core. It is a manifestation of energy's stabilizing nature, confining active energy (quarks and gluons) within the nucleus to maintain order.

The Weak Nuclear Force: The weak force governs processes like radioactive decay and neutrino interactions. It allows particles to transition between different states, facilitating the transformation of one type of matter into another. This force reflects the dynamic aspect of energy transitions, enabling changes that drive phenomena such as the fusion reactions powering stars.

Quantum Stability and Transition: Both nuclear forces demonstrate how energy transitions govern stability and transformation at the quantum level. They ensure that energy, whether stable or active, is balanced within the confines of the atom.

How These Forces Connect with the Universal Energy Pool

Electromagnetic and nuclear forces, though seemingly confined to the quantum and atomic realms, are integral components of the broader energy dynamics within the infinite energy pool.

Interactions with the Stable Energy Medium: Both forces arise from and interact with the stable energy reservoir, drawing from it to fuel dynamic processes and returning to it during stabilization. For instance, the emission of light during electron transitions reflects active energy interacting with the stable medium, releasing energy while maintaining equilibrium.

Energy Transitions Across Scales: The behavior of these forces reveals the continuity of energy transitions across scales. The same principles that govern nuclear stability also influence larger systems, such as electromagnetic radiation propagating through interstellar space.

Cosmic Implications: At cosmic scales, the energy dynamics of electromagnetic and nuclear forces influence phenomena like star formation, supernovae, and the emission of gamma rays. These forces bridge the quantum processes within stars with the macroscopic events observable across the universe.

Unified Perspective on Forces

When viewed through UED, electromagnetic and nuclear forces are not isolated phenomena but integral expressions of energy's dual nature:

Electromagnetic Force: Active energy in motion, shaping interactions across scales and connecting systems through its universal reach.

Strong and Weak Nuclear Forces: Stabilizers of energy at the quantum scale, ensuring coherence within matter and driving essential transformations.

Together, these forces exemplify the principles of stability and dynamism, operating within the infinite energy pool to maintain universal harmony. By understanding these forces as dynamic energy interactions, we reveal the interconnectedness of all phenomena, from the subatomic to the cosmic, and reinforce the unity of the universe's energetic fabric.

4.5 Unified Force: The Illusion of Separation

In our understanding of forces, classical frameworks often categorize them into distinct phenomena: gravity, electromagnetism, and the nuclear forces. While these categorizations help describe the interactions governing different aspects of the universe, they can obscure a deeper truth: all forces are unified expressions of a single underlying energy principle. The distinctions we observe arise not from fundamentally different origins but from localized manifestations of energy transitions. By exploring this unifying thread, we can move beyond the illusion of separation and uncover the interconnected nature of forces within the framework of Unified Energy Dynamics (UED).

All Forces as Facets of a Single Underlying Energy Principle

Energy as the Common Source: Forces are emergent phenomena arising from the dynamic interplay between stable and active energy within the infinite energy pool. Gravity, electromagnetic forces, and nuclear interactions are not independent but represent energy transitions tailored to specific contexts, such as the behavior of mass, charge, or subatomic particles.

Unified Expression across Scales: At the quantum level, nuclear forces stabilize atomic nuclei by managing energy transitions within incredibly small distances. At larger scales, electromagnetic forces govern interactions between charged particles and propagate energy through waves. Gravity, often seen as distinct due to its weak nature, emerges from interactions between mass and the stable energy medium, creating an overarching framework for large-scale coherence.

Harmony within Diversity: Though forces manifest differently, they are interconnected, each contributing to the balance of energy dynamics within the cosmos. **For example**, the interplay between gravitational and electromagnetic forces is essential for star formation, where the collapse of matter (governed by gravity) triggers nuclear fusion (powered by electromagnetic and nuclear forces).

How Apparent Distinctions Arise from Localized Energy Expressions

Contextual Manifestations: Forces appear distinct because they operate under specific conditions that emphasize particular aspects of energy dynamics: **Gravity** is prominent in systems with significant mass and scale, where the stable energy medium influences motion and structure. **Electromagnetic Forces** dominate interactions involving charged particles, where active energy manifests as radiation or attraction and repulsion. **Nuclear Forces** come into play in the confined spaces of atomic nuclei, where stability is critical to overcoming repulsion between protons.

Localized Energy Effects: The stable energy pool acts as the backdrop for all phenomena, but localized disturbances within this medium give rise to the forces we observe:

✓ Energy interacting with charge fields creates electromagnetic waves.

✓ Energy confined within the nucleus creates the strong and weak forces.

✓ The curvature of the stable energy medium around mass creates gravitational effects.

Observer Perception and Scale: Human understanding often separates forces based on observable phenomena at specific scales, such as atomic interactions versus celestial mechanics. This separation is a product of perception, not a reflection of distinct origins. In reality, all forces stem from the same energy dynamics, differentiated only by the conditions under which they manifest.

Exploring the Unifying Thread Across Diverse Interactions

From Gravity to Quantum Forces: The apparent difference between gravity and other forces arises from the unique role of the stable energy medium: Gravity reflects how mass interacts with the stable reservoir, creating curvature and motion. Electromagnetic and nuclear forces arise from localized active energy transitions, operating within the stable framework.

Energy's Dual Nature: The dual states of energy—stable and active—are the unifying thread connecting all forces: Stable energy provides the coherence that underpins gravitational stability and nuclear binding. Whereas, Active energy drives transformations, powering electromagnetic radiation and quantum interactions.

Bridging Scales and Domains: By understanding forces as energy expressions, we can connect phenomena across scales:

✓ Quantum entanglement and wave-particle duality are micro-level manifestations of energy dynamics.

✓ Galactic motion and gravitational waves are macro-level manifestations governed by the same principles.

Toward a Unified Vision of Forces

The illusion of separation among forces dissolves when viewed through the lens of UED. Gravity, electromagnetic forces, and nuclear interactions are not disparate entities but facets of a single, unified energy principle. Their distinctions are contextual, arising from the conditions under which energy transitions occur.

By embracing this unified vision, we gain a deeper understanding of the interconnectedness of all phenomena:

✓ **Unifying Medium**: The infinite energy pool serves as the common origin and framework for all forces.

✓ **Dynamic Interplay**: The transitions between stable and active energy manifest as the forces driving motion, interaction, and transformation.

✓ **Universal Harmony**: The coherence of the cosmos arises from this unity, ensuring balance and continuity across scales.

This perspective not only reconciles the diversity of forces but also invites new explorations into the fundamental nature of reality, where energy stands as the silent architect of the universe's intricate and harmonious design.

4.6 Forces and Cosmic Harmony

Forces are not merely mechanical phenomena; they are the universe's architects, shaping its structure, enabling transformation, and maintaining equilibrium across scales. Within the framework of Unified Energy Dynamics (UED), forces are revealed as emergent manifestations of energy interactions, bridging stability and dynamism to sustain the universe's intricate balance. This section explores how forces maintain cosmic harmony, act as agents of structure, and drive transformation across the vast expanse of reality.

How Forces Maintain Equilibrium in the Universe

Balance Between Stability and Motion: Forces emerge as mechanisms that mediate the interaction between stable and active energy, ensuring equilibrium: **Gravity** balances the attraction of masses with their dynamic motion, creating orbital systems that maintain stability over astronomical timescales. **Electromagnetic Forces** regulate interactions between charged particles, maintaining the structural integrity of atoms and molecules.

Universal Regulation: Forces act as self-regulating phenomena within the energy pool: In galaxies, gravity counterbalances centrifugal forces generated by rotational motion, preventing disintegration while sustaining cosmic architecture. At the quantum scale, the nuclear forces prevent atomic nuclei from flying apart due to proton repulsion, ensuring the continuity of matter.

Dynamic Equilibrium: Forces are not static; they adapt to changes in energy distribution, maintaining harmony in evolving systems: A collapsing star transitions into a supernova due to the interplay of gravitational and nuclear forces, redistributing energy and matter while preserving the system's balance. Tidal forces in celestial systems dynamically adjust to shifts in orbital parameters, sustaining the interconnected dance of celestial bodies.

The Interplay of Stability and Dynamism in Force Interactions

Forces as Mediators of Energy Transitions: Forces operate at the interface of stable and active energy, converting potential energy into motion and transformation: **Electromagnetic Waves** propagate energy through space, bridging the gap between stability and dynamism. **Gravitational Waves** reflect disturbances in the stable energy medium, spreading the effects of massive cosmic events.

Local and Global Interactions: Forces influence systems at both localized and universal scales: In atoms, electromagnetic and nuclear forces maintain structure and enable interactions that fuel life processes. On a cosmic scale, gravity organizes galaxies, enabling the clustering of matter while accommodating expansion driven by active energy transitions.

Sustaining Motion and Transformation: The dynamic interplay of forces ensures that motion and transformation occur within a stable framework: In planetary systems, gravitational forces create stable orbits, allowing for the evolution of life and ecosystems. In stars, nuclear forces drive fusion, transforming hydrogen into heavier elements while releasing energy that sustains the stellar lifecycle.

Forces as the Architects of Structure and Transformation

Shaping the Universe's Architecture: Forces construct and sustain the structural framework of the cosmos: Gravity organizes matter into galaxies, stars, and planets, defining the large-scale structure of the universe. Electromagnetic forces shape the atomic and molecular bonds that form the foundation of matter, chemistry, and biology.

Driving Transformation: Forces are agents of change, enabling the evolution and reconfiguration of systems: In stellar formation, gravitational collapse initiates nuclear fusion, transforming gas clouds into radiant stars. In chemical reactions, electromagnetic forces rearrange atoms, giving rise to the diversity of matter and the complexity of life.

The Cycle of Creation and Dissolution: Forces guide the perpetual cycle of formation and dissolution: Black holes exemplify this cycle, where gravity consumes matter, releasing active energy into the stable energy pool and maintaining universal equilibrium. Supernovae distribute elements across the cosmos, enabling the formation of new stars and planetary systems.

Forces as Agents of Cosmic Harmony

Forces embody the universe's innate drive toward balance and interconnectedness. They mediate the dynamic relationship between stability and motion, ensuring that the cosmos evolves cohesively:

✓ **Gravity** binds masses into coherent systems while accommodating the motion of celestial bodies.

✓ **Electromagnetic Forces** govern interactions at atomic and molecular levels, enabling the complexity of life and matter.

✓ **Nuclear Forces** stabilize the subatomic realm, maintaining the integrity of matter while enabling transformative processes like fusion.

This unified perspective reveals forces as more than mechanical entities; they are the architects of harmony, sustaining the universe's intricate balance while driving its perpetual evolution. Through their interactions, the cosmos achieves both structure and transformation, reflecting the dynamic interplay of energy that underpins all existence.

In summary, Forces, when examined through the lens of Unified Energy Dynamics (UED), reveal themselves as more than mechanical or physical entities; they emerge as the tangible expressions of energy's perpetual flow and interaction. These forces, which shape and sustain the universe, are neither isolated phenomena nor independent concepts. Instead, they are unified manifestations of the same underlying principle: energy in action. This conclusion not only reaffirms the cohesive nature of all forces but also bridges the realms of science and metaphysics, offering a profound framework for understanding reality.

4.7 Reaffirming the Unified Nature of All Forces

A Singular Foundation: All forces—gravitational, electromagnetic, nuclear, or hypothetical—are facets of a singular energy framework: **Gravity** emerges as an interaction between mass and the stable energy medium, demonstrating the binding role of energy at large scales. **Electromagnetic and Nuclear Forces** operate as dynamic expressions of active energy, stabilizing matter at microscopic levels and enabling transformation. Apparent distinctions between forces are localized phenomena, arising from specific interactions and contexts within the energy pool.

A Continuum Across Scales: Forces operate seamlessly across quantum and cosmic scales, driven by the interplay of active and stable energy: At the **quantum level**, forces govern particle interactions, shaping the behavior of matter and energy at its smallest scales. At the **cosmic scale**, forces maintain the architecture of galaxies, stars, and planetary systems, ensuring the coherence of the universe.

Unifying Energy Principles: By understanding forces as emergent phenomena of energy transitions, we move beyond fragmented scientific frameworks: Forces are not separate entities but expressions of the same underlying energy medium, bound together by principles of stability and dynamism. This perspective integrates the diversity of forces into a cohesive narrative, revealing their interconnectedness and shared origins.

Forces as Tangible Expressions of Energy's Flow

Energy in Motion: Forces embody the dynamic aspects of energy, translating its transitions into observable effects: **Active Energy** drives forces like electromagnetism and nuclear interactions, enabling motion, transformation, and creation. **Stable Energy** underpins gravitational interactions, providing the framework for continuity and order.

Energy as a Creative and Stabilizing Force: Forces facilitate the cycle of creation, transformation, and dissolution:

✓ **Gravitational Forces** cluster matter into stars and galaxies, creating the structures of the universe.

✓ **Electromagnetic Forces** enable chemical bonds and life processes, shaping the complexity of existence.

✓ **Nuclear Forces** stabilize atomic nuclei while driving transformative processes like fusion and decay.

Observable Interactions: Forces translate the invisible flow of energy into the tangible interactions we observe: They sustain the stability of matter while fueling the motion and dynamics of celestial and subatomic systems. Through forces, energy manifests in ways that connect the micro and macro realms, offering insights into the universe's behavior.

Bridging Science and Metaphysics Through a Cohesive Framework

A Scientific and Philosophical Union: The unified nature of forces offers a bridge between empirical science and metaphysical inquiry: Forces, as expressions of energy, connect the measurable (motion, interaction) with the immeasurable (stability, balance). This dual perspective enriches our understanding of existence, allowing us to explore both physical phenomena and their deeper implications.

A New Paradigm for Exploration: By viewing forces as energy in action, we transcend the limitations of classical and modern physics: The traditional division between forces dissolves, replaced by a holistic understanding rooted in energy dynamics. This perspective invites interdisciplinary exploration, fostering dialogue between physics, philosophy, and metaphysics.

Energy as the Universal Connector: Forces illustrate energy's role as the unifying thread that binds the cosmos: They highlight energy's capacity to shape, sustain, and transform reality across all scales. This understanding provides a cohesive framework for exploring the universe, one that is both scientifically rigorous and philosophically profound.

A Unified Vision of Forces

Forces are not fragmented or isolated phenomena but expressions of a single, unified energy medium. They mediate the interactions between active and stable energy, translating transitions into the motion, structure, and transformation we observe. From the gravitational pull of planets to the quantum binding of particles, forces sustain the universe's intricate harmony, ensuring its perpetual evolution.

By embracing this unified perspective, we bridge the gaps between scientific observation and metaphysical understanding. Forces, as tangible manifestations of energy in action, reveal a universe that is not only interconnected but also profoundly dynamic—a cosmos driven by the eternal flow of energy, guided by the principles of stability and dynamism, and held together by the cohesive framework of Unified Energy Dynamics.

Summary: The Nature of Forces

This chapter delves into the nature of forces, redefining them as emergent phenomena within the **Unified Energy Dynamics (UED)** framework. Forces are presented not as isolated entities but as interconnected expressions of energy transitions, shaped by the interplay between stability and dynamism in the universal energy pool.

1. Forces as Expressions of Energy

Forces arise from the dual nature of energy:

- **Stable Energy**: Provides the structural coherence necessary for forces to emerge and propagate.

- **Active Energy**: Drives interactions and transformations, manifesting as observable forces.

Forces are described as the "language" through which energy transitions influence matter and systems, unifying microscopic and macroscopic phenomena.

2. Redefining Forces in the UED Framework

Gravity: Traditionally viewed as the attraction between masses, gravity is redefined as an interaction between mass and the stable energy medium. This interaction shapes spacetime, creating gradients that manifest as gravitational effects.

Electromagnetic Forces: These forces arise from the interaction of charged particles within the energy medium. They govern atomic structures, light propagation, and energy transfer, highlighting their dynamic and universal reach.

Nuclear Forces:

- **Strong Nuclear Force** binds protons and neutrons within atomic nuclei, ensuring stability.

- **Weak Nuclear Force** drives processes like radioactive decay and particle transformations, showcasing energy's capacity for change.

3. Forces as Unified Phenomena

The distinctions between forces (gravitational, electromagnetic, and nuclear) are contextual, arising from localized energy interactions.

All forces originate from the same energy pool, demonstrating a unified principle:

✓ Gravity dominates at macroscopic scales.
✓ Electromagnetic forces influence atomic and molecular systems.
✓ Nuclear forces operate within atomic nuclei.

4. Observable Effects across Scales

Quantum Scale: Forces mediate particle interactions, shaping phenomena like quantum entanglement and wave-particle duality.

Cosmic Scale: Forces govern the motion of galaxies, the behavior of stars, and the dynamics of planetary systems.

5. Toward a Unified Vision

The illusion of separation among forces dissolves in the UED framework, where all forces are facets of energy dynamics.

By understanding forces as emergent phenomena, the chapter bridges traditional physics with a holistic view, emphasizing the interconnectedness of all phenomena.

Conclusion

This chapter redefines forces as interconnected expressions of energy within the infinite energy medium. By integrating stable and active energy dynamics, the chapter provides a unified perspective on the behavior of forces across scales, from quantum interactions to cosmic architecture.

Chapter 5: The First Disturbance

From where did it all begin? This timeless question has fascinated humanity across ages, inspiring myths, philosophies, and groundbreaking scientific theories. The prevailing Big Bang theory, though widely accepted, envisions the universe's birth as a singular explosive event—a beginning that echoes across the cosmos. But what if the answer lies deeper? What if the origin of all existence was not an explosion but a gentle ripple—*The First Disturbance*?

Building on the foundation of stable and active energy dynamics, this chapter presents an alternative to the Big Bang—a quantum event within the infinite stable energy pool. This event did not happen once; it is an eternal phenomenon, continuously reshaping the universe at the quantum level. *The First Disturbance* marks the moment when absolute stability is nudged into motion, birthing active energy, creating photons as carriers of this transformation, and initiating the cycles of creation and evolution.

This chapter is not just an exploration of beginnings; it is an invitation to reimagine the origin of reality. Through *The First Disturbance,* we will uncover the processes that govern energy, motion, and consciousness, offering a compelling vision of how the universe emerges not from chaos but from the quiet awakening of infinite potential.

5.1 Introduction: What Is the First Disturbance?

At the core of existence lies a profound question: what initiated it all? Was there a single event, a moment in time, or an eternal process that brought the universe into being? While traditional cosmology attributes this genesis to the Big Bang—a massive explosion marking the birth of time, space, and matter—Unified Energy Dynamics offers a different perspective.

The *First Disturbance* redefines this moment of origin. It was not a chaotic explosion but a subtle, quantum event arising within the infinite stable energy pool. This disturbance—like a ripple across a perfectly calm lake—was the first interaction, the first shift from stability to motion. It was the genesis of active energy, the spark that set the cycles of creation and dissolution into motion.

Quantum superposition, the phenomenon where multiple potential states coexist, emerges as the root of this disturbance. Within the stable energy pool, a localized imbalance triggered an awakening—a transition from potential to actual, birthing energy that would transform into matter, radiation, and the structures of the cosmos. This process is neither confined to a single moment nor a one-time event; it is a continuous, infinite dance of creation unfolding at the quantum level.

The *First Disturbance* is not merely an origin story; it is the framework through which we understand the dynamic interplay of energy and existence. By exploring its role, we uncover the processes that sustain the universe, challenge conventional cosmological narratives, and illuminate the profound interconnectedness of reality.

5.2 The Birth of Consciousness: Energy in Transit

The *First Disturbance* was more than the birth of motion and matter—it marked the emergence of consciousness itself. In the infinite, stable energy pool, the transition initiated by the disturbance not only created active energy but also gave rise to the mechanism that connects it back to its source. This mechanism, this bridge, is what we recognize as consciousness: energy in transit.

Consciousness is not merely a trait of living beings but a universal phenomenon born from the dynamic interplay between stable and active energy. It is the bridge that allows active energy to interact with, influence, and ultimately return to the stable energy pool. This cycle of transit and interaction is the foundation of all existence, from the motion of subatomic particles to the evolution of galaxies and life itself.

In living beings, consciousness takes on an intellectual and perceptual form, shaped by the active energy that animates their physical systems. The mind, as a functional manifestation of this energy, allows humans to observe, analyze, and create. But at its core, consciousness is universal, present even at the quantum level. The behavior of particles, their superposition, entanglement, and wave-particle duality, reflects this underlying awareness—energy recognizing and responding within the infinite pool.

The *First Disturbance* did not merely spark the cosmos; it awakened the interconnected dance of energy and existence. Consciousness emerged as the medium through which this dance unfolds, bridging the stable and the active, the infinite and the finite. It is both the observer and the observed, the conveyor of the universe's story and the essence of its profound unity.

5.3 Photon: The Tangible Form of Universal Energy

Photon as a Visible and Detectable Manifestation of Energy

Within the UED framework, photons emerge as the most tangible and observable form of the *Universal Energy Pool*, which includes stable energy (also referred to in modern science as *dark energy*). Photons are not just carriers of electromagnetic radiation; they serve as detectable signatures of the interactions and transitions occurring within this vast and infinite energy medium. Every instance of photon emergence—whether in friction, atomic transitions, or cosmic-scale events—represents the energy pool's dynamic response to disturbances.

Photon as Proof of Energy's Duality

The photon's dual nature, exhibiting both particle-like and wave-like behavior, embodies the fundamental duality of universal energy: **stability and dynamism**.

Wave Nature: The wave aspect of the photon reflects the infinite and continuous presence of the stable energy pool. This is the unseen, omnipresent foundation that sustains the motion and propagation of electromagnetic waves across the cosmos.

Particle Nature: The particle nature, or the discrete energy packets (quanta) we observe as photons, symbolizes the active energy manifesting locally and momentarily within the universal energy medium. It is energy in action, driving observable phenomena and interactions.

This duality mirrors the broader interplay between stable and active energy: the infinite pool of stability and the localized bursts of dynamism. The photon becomes a microcosm of this universal principle, a visible clue to the hidden mechanics of existence.

Photon's Emergence and Dissolution in UED

The lifecycle of a photon offers direct insight into the behavior of universal energy:

Emergence: Photons arise whenever energy transitions from one form to another. They emerge as sparks of light during friction, heat radiation, electron transitions, and cosmic events like nuclear fusion or black hole evaporation. Their emergence marks the transformation of active energy into a detectable state.

Propagation: Medium independence is a defining feature of photons. Unlike waves needing a material medium, photons traverse the stable energy pool itself, making them universal conveyors of energy across vast distances.

Dissolution: Photons do not vanish; they dissolve back into the stable energy pool, maintaining the conservation of energy. This process is vividly observed at black holes, where photons seemingly disappear as they merge back into the infinite reservoir of stability.

Photons as a Reflection of Universal Energy's Eternal Dance

Every photon is a reminder of the *First Disturbance*—the primal quantum superposition that initiated existence. Each photon that emerges is an echo of that original event, representing the continuous creation, motion, and eventual return to stability. By existing simultaneously as wave and particle, photons reveal the harmonious tension between the infinite stillness of stable energy and the ceaseless action of active energy.

Photon and Consciousness: A Deeper Connection

In the UED perspective, photons are more than just particles of light—they are manifestations of the universal energy pool in transit. This transit, much like consciousness, bridges the stable reservoir of potential with the dynamic expressions of reality. Just as consciousness allows humans to interact with and perceive existence, photons enable the universe to transfer energy, information, and motion across its vast expanse.

Implications of Photon Duality in UED

Revisiting Fundamental Physics: The photon's duality is not an isolated quantum phenomenon but a direct reflection of the dual states of universal energy, providing a unified explanation for light's behavior.

Probing Dark Energy: If photons are localized expressions of the universal energy pool, their study offers a pathway to understanding *dark energy*, the enigmatic force that permeates the cosmos.

Connection to the First Disturbance: Photons serve as messengers of the primal event, carrying the signature of the quantum superposition that began all existence.

By redefining photons as the visible, detectable form of universal energy, the UED framework not only clarifies their nature but also elevates their role in the grand tapestry of existence. They are the carriers of energy's eternal dance—a dance that began with the First Disturbance and continues in every transformation, every beam of light, and every quantum event.

5.4 Quantum Superposition: The Foundation of Disturbance

How Superposition Creates Fluctuations in the Stable Energy Pool

Quantum superposition lies at the heart of the *First Disturbance*, serving as the primal mechanism that initiates fluctuations within the stable energy pool. In its simplest form, superposition is the simultaneous existence of multiple potential states, creating a state of inherent uncertainty.

In the context of the stable energy pool, this quantum phenomenon triggers localized disturbances—moments where energy transitions from a state of absolute stability to active dynamism. These fluctuations manifest as the emergence of active energy, breaking the stillness of the stable energy pool.

The Catalyst of Change: Superposition acts as the spark that ignites energy transformations, disrupting the homogeneity of the stable energy medium. Each disturbance creates an opportunity for the formation of observable phenomena, from particles at the quantum scale to galaxies at the cosmic level.

Localized Events, Universal Impact: Though quantum superposition operates at minute scales, its effects ripple across the universe, driving the perpetual motion of energy and matter. These fluctuations are not isolated but part of a continuous process, shaping the dynamic equilibrium of the cosmos.

The Continuous Process of Energy Transformation at Quantum Levels

Superposition is not a one-time event; it is an ongoing process that sustains the universe's dynamic nature. At quantum levels, this phenomenon facilitates the continuous exchange between stable and active energy, maintaining the cycle of creation, transformation, and dissolution.

The Birth of Particles: Energy disturbances at the quantum level lead to the emergence of particles, embodying active energy localized within the stable energy pool. These particles exhibit dual behaviors, such as wave-particle duality, reflecting their transient existence within the energy medium.

The Driver of Interactions: Quantum superposition enables particles to interact with one another and with the stable energy pool, creating phenomena like entanglement and energy transfer. This interaction is the foundation of all observable forces, from electromagnetic interactions to gravitational waves.

The Universal Engine of Motion: Superposition ensures that energy transformations occur continuously, driving the motion of particles, the behavior of matter, and the evolution of systems across scales. This process underpins the universe's unending activity, sustaining the balance between stability and dynamism.

Superposition: The Seed of Reality

Quantum superposition, as the foundation of disturbance, is more than a quantum phenomenon—it is the engine of existence itself. By enabling fluctuations in the stable energy pool, it creates the conditions necessary for energy to transition, transform, and manifest as the observable universe. Every particle, wave, and photon carries the signature of this foundational process, echoing the primal *First Disturbance* that set the cosmos into motion.

Through this lens, superposition is not merely a mathematical abstraction but a profound principle that bridges the quantum and cosmic, the transient and the eternal, making it the cornerstone of the universe's perpetual evolution.

5.5 Photon-Matter Interactions: Evidence of the First Disturbance

Examining Phenomena Like the Photoelectric Effect

Photon-matter interactions provide compelling evidence of the *First Disturbance* and the ongoing dynamic exchange between stable and active energy. Among the most striking examples is the **photoelectric effect**, where photons striking a material cause the ejection of electrons. This phenomenon illustrates the transformative power of photons as vehicles of energy and their capacity to initiate observable change.

The Nature of the Photoelectric Effect: When light photons strike the surface of certain materials, the energy they carry is transferred to electrons within the material, ejecting them from their atomic orbits. This process demonstrates how photons act as intermediaries, carrying energy derived from the stable energy pool and delivering it to matter.

Evidence of Quantized Energy: The photoelectric effect reveals the quantized nature of photons, showcasing their discrete energy packets that correspond to the frequency of light. This quantization underscores the photon's role as a fundamental unit of active energy transition, reflecting the localized disturbances in the stable energy pool.

Validation of Energy Dynamics: The interaction between photons and electrons validates the principles of energy transformation, showing how energy transitions drive both micro-level and macro-level phenomena. This process echoes the universal dynamic initiated by the *First Disturbance*, connecting light, matter, and motion.

How Photons Interact with Matter to Reveal Energy Transitions

Beyond the photoelectric effect, photons interact with matter in numerous ways, providing windows into the dynamic nature of energy. These interactions illustrate the continuous interplay between the active energy carried by photons and the stable energy pool.

Absorption and Emission: Photons are absorbed by matter, transferring their energy to excite electrons to higher energy states. Conversely, as excited electrons return to lower energy states, photons are emitted, illustrating the cyclical nature of energy transitions.

Reflection and Refraction: Photons encountering surfaces can reflect or refract, revealing how light interacts with the structural properties of matter. These interactions are governed by the properties of the energy medium, demonstrating its universal influence.

Fluorescence and Phosphorescence: In fluorescence, absorbed photons are re-emitted almost instantly, while in phosphorescence, the re-emission is delayed. These phenomena highlight the role of matter in storing and releasing energy derived from photons, showcasing energy's persistence across forms.

Energy Transfer in Life Processes: In biological systems, photons power photosynthesis, where light energy is converted into chemical energy, sustaining life on Earth. This interaction underscores the universal importance of photon-mediated energy transitions in shaping existence.

Photon Interactions: A Legacy of the First Disturbance

Every photon-matter interaction is a direct legacy of the *First Disturbance*, revealing the intricate processes through which energy transitions shape the universe. Photons serve as carriers of the initial disturbance, delivering its influence across time and space.

By observing how photons interact with matter, we uncover the hidden dynamics of energy transitions, bridging the quantum and macroscopic realms. These interactions offer profound insight into the nature of the universe, affirming that the *First Disturbance* continues to resonate through every light particle, every energy exchange, and every transformation in existence.

5.6 The Universal Energy Pool: Source and Sink

The Role of the Stable Energy Reservoir in Birthing and Absorbing Photons

The **Universal Energy Pool**, described as the infinite stable reservoir, acts as both the **source** and **sink** for photons, the visible manifestations of active energy transitions. This dynamic underscores the dual role of the stable energy pool in creating and maintaining universal harmony.

Photons as Energy Emissaries: Photons emerge from the energy pool during transitions, such as matter interactions or energy transformations, acting as emissaries of the universal energy dynamics. **For example**, photons are released when electrons transition between energy states or during thermodynamic processes like heat radiation. This emergence reflects the active energy momentarily released from the stable energy pool, bridging the quantum and macroscopic realms.

Absorption and Return: Just as photons emerge from the stable energy reservoir, they are absorbed back into it after delivering their energy. This cyclical process ensures the conservation and coherence of energy across all scales, maintaining the balance between creation and dissolution. The disappearance of photons in phenomena like black hole event horizons illustrates this absorption, where active energy seamlessly integrates back into stability.

Dynamic Equilibrium: The stable energy pool's dual role as a source and sink highlights its ability to sustain the dynamic equilibrium of the cosmos. It enables energy to manifest as active phenomena (e.g., light and heat) and subsequently stabilizes these phenomena by reabsorbing their energy into the infinite reservoir.

How the First Disturbance Sustains Cosmic Dynamics

The *First Disturbance*, initiated by quantum superposition, set in motion the perpetual cycles of energy transformation. This primal event established the rules of the universal energy pool, enabling the emergence of photons and the continual sustenance of cosmic processes.

Catalyzing Energy Fluctuations: The *First Disturbance* created localized fluctuations in the stable energy pool, generating the dynamic interplay between stability and motion. These fluctuations gave rise to photons as the first messengers of energy transitions, propagating the influence of the disturbance across the universe.

Sustaining Cosmic Motion: Photons, as products of the *First Disturbance*, facilitate energy transitions that drive motion, transformation, and evolution across the cosmos. From the propagation of light through space to the transfer of heat and radiation, these processes reflect the sustained influence of the initial event.

Connecting the Micro and Macro Realms: The continuous generation and absorption of photons reveal the seamless connection between the quantum scale (where energy transitions occur) and the macroscopic universe. This interplay ensures that the principles initiated by the *First Disturbance* resonate through every layer of reality, maintaining cosmic harmony.

The Stable Energy Pool as the Pillar of Existence

The stable energy reservoir is not a passive backdrop but the **pillar of existence**, orchestrating the flow of energy across the universe. It serves as the wellspring from which active energy emerges and the sink into which it dissolves, perpetuating the cycles of creation and transformation.

Through the lens of Unified Energy Dynamics, we understand the energy pool as the silent architect of the cosmos—a boundless medium that ensures balance, sustains motion, and connects all phenomena to the primordial spark of the *First Disturbance*.

5.7 The Eternal Dance: Creation, Emission, and Return

The Cyclical Process of Energy's Emergence and Dissolution

At the heart of the universe lies a perpetual cycle of **creation, emission, and return**, an eternal dance of energy transitioning between its stable and active states. This process, initiated by the *First Disturbance*, sustains the dynamic equilibrium of the cosmos and ensures the continuity of existence.

Emergence of Energy: Disturbances within the stable energy pool create active energy phenomena. Photons, the primary emissaries of energy transitions, emerge from this reservoir during processes like friction, heat radiation, and nuclear fusion. Each emergence represents energy transitioning from a latent, stable state to a dynamic, observable form, fueling motion, transformation, and interaction.

Emission and Propagation: Once emitted, photons propagate through space as carriers of energy and information. These emissions power cosmic and terrestrial phenomena, from the light of stars illuminating galaxies to the heat radiating from a campfire. The motion of photons, independent of traditional mediums, reflects their inherent connection to the energy pool, serving as bridges across vast scales.

Return to Stability: The cycle completes as photons are absorbed back into the stable energy pool, dissolving their active energy into stability. This return occurs during interactions with matter, such as when light is absorbed by surfaces or disappears into black holes, illustrating the seamless reintegration of energy. Through this process, the universe maintains its coherence and conserves energy across all systems.

Photons as Symbols of the Universe's Dynamic Equilibrium

Photons, with their dual nature as particles and waves, embody the **dynamic balance** between stability and motion. Their behavior exemplifies the intricate interplay between creation and dissolution, highlighting their role as symbols of the universe's equilibrium.

Agents of Creation and Transformation: Photons catalyze countless transformations, from photosynthesis in plants to the triggering of chemical reactions. They act as the visible threads of the cosmic fabric, connecting phenomena across the quantum and cosmic scales.

Markers of Energy's Cycles: The constant emergence and disappearance of photons reflect the cyclical nature of energy dynamics. Their journey—from creation to emission and eventual dissolution—serves as a microcosm of the universe's larger cycles of birth, transformation, and return.

Bridging Stability and Motion: Photons illustrate how energy, while rooted in stability, can manifest as dynamic phenomena. This duality reaffirms the principles of Unified Energy Dynamics, where stability and motion coexist to sustain the cosmos.

The Eternal Dance of Energy

The universe's harmony is built on this eternal dance of creation, emission, and return—a process that embodies the essence of existence. Through photons, we glimpse the intricate balance between the stable reservoir and the dynamic force of active energy.

This cyclical interplay not only sustains the universe but also unites its vast complexity into a cohesive whole. By understanding this dance, we move closer to unveiling the profound simplicity underlying the cosmos, where every photon tells the story of the universe's perpetual motion and its infinite equilibrium.

5.8 Implications of the First Disturbance

Reimagining the Origins of the Universe beyond the Big Bang

The *Big Bang*, long considered the definitive starting point of the universe, presumes the existence of matter—a paradox that leaves the question of its origins unanswered. The **First Disturbance**, as conceptualized in Unified Energy Dynamics (UED), offers a transformative perspective. Instead of a singular explosive event marking the birth of the cosmos, the First Disturbance represents an ongoing process of **energy transitioning into matter** at the quantum level.

Energy to Matter: The True Genesis

The First Disturbance initiated the conversion of stable energy into active energy, giving rise to matter.

This quantum-level transformation occurred in localized pockets of the stable energy pool, creating the building blocks of physical reality.

Unlike the instantaneous creation implied by the Big Bang, this process unfolds gradually and continuously, with matter emerging as disturbances propagate through the energy medium.

Big Bang as a Secondary Phenomenon

The Big Bang, rather than being the origin, can be understood as a **cosmic collision of matter** that caused a large-scale disturbance in the stability of the energy pool.

This event accelerated the processes of creation and energy transitions, amplifying the distribution of matter and energy throughout the universe.

It highlights the dynamic interplay between stability and motion, where localized disruptions cascade into cosmic phenomena.

Rethinking Temporal Beginnings

The concept of the First Disturbance eliminates the need for a temporal "beginning," as the energy pool exists infinitely, outside the constraints of time.

The continual emergence and dissolution of matter reflect a timeless cycle of transformation, rooted in the interplay of energy dynamics.

This view shifts the focus from a singular event to the eternal processes that sustain existence.

The First Disturbance as the Continual Process of Creation

The First Disturbance is not a singular event frozen in the past but an ongoing mechanism that drives the universe's evolution. It establishes the foundation for all observable phenomena by perpetuating the interplay between energy and matter.

Creation at the Quantum Level: Energy disturbances at the quantum scale continually generate particles, demonstrating the First Disturbance's ongoing role in matter creation. This process mirrors the emergence of photons during energy transformations, highlighting the seamless connection between quantum and cosmic scales.

Cosmic Evolution Through Disturbance: As matter accumulates and interacts, larger-scale disturbances shape galaxies, stars, and planets. The same principles that govern quantum fluctuations drive the formation of complex structures, uniting the micro and macro realms under a single framework.

Perpetual Creativity of the Universe: The First Disturbance exemplifies the universe's inherent creativity, where energy transitions fuel the continual birth of matter and phenomena. This perpetual process ensures the dynamic equilibrium of the cosmos, balancing creation and dissolution to sustain universal harmony.

A Transformative Perspective on Origins

By reimagining the origins of the universe through the lens of the First Disturbance, we uncover a vision of reality that is both timeless and dynamic. It moves beyond the limitations of the Big Bang theory, embracing a framework where **creation is not a singular event but an eternal process**. This perspective aligns the origins of existence with the principles of Unified Energy Dynamics, offering profound insights into the interplay of energy, matter, and the cosmos.

5.9 Conclusion: The Spark That Connects All

Reaffirming the First Disturbance as the Origin of Consciousness and Energy Flow

The **First Disturbance**, born out of quantum superposition within the infinite energy pool, stands as the primal event that initiated the eternal dance of creation, transformation, and dissolution. It is not merely a theoretical construct but the foundation of all existence—a process that continuously connects stable and active energy, birthing both matter and consciousness.

This phenomenon defines the flow of energy throughout the universe, from the smallest quantum events to the grandeur of cosmic evolution. The First Disturbance is also the origin of **consciousness**, the dynamic bridge between stability and motion. As energy transitions occur, consciousness emerges as the interplay of active energy and the stable reservoir, shaping not only matter but the ability to perceive and interact with the cosmos.

Photons as Conveyors of Existence in an Infinite Universe

Photons, the first messengers of the First Disturbance, exemplify the duality and dynamism of the universal energy pool. They emerge whenever energy transitions occur, carrying information and vitality across the expanse of the cosmos. These radiant particles are not mere carriers of electromagnetic waves but tangible expressions of the energy pool itself—visible, measurable, and deeply intertwined with the essence of existence.

As photons traverse the infinite universe, they deliver the spark of energy to sustain life, light stars, and fuel the perpetual cycle of creation. Their disappearance back into the stable energy pool during phenomena like black hole formation underscores their role in maintaining the dynamic equilibrium of the cosmos.

The Spark That Unites All

The First Disturbance is the universal **spark** that connects all phenomena, bridging the quantum and the cosmic, the physical and the metaphysical. It reveals a universe that is not static but an infinite interplay of energy transitions, with photons serving as the conduits of this dynamic flow. By understanding the First Disturbance, we move closer to uncovering the profound interconnectedness of existence—a reality where creation is not an event of the past but an eternal process, continuously unfolding across the boundless energy medium of the universe.

Summary of Part I

Traditional Approach vs. Unified Energy Dynamics (UED)

Part I of this book establishes a revolutionary perspective on the nature of energy, its dynamics, and its universal implications, contrasting the **traditional scientific approach** with the innovative framework of **Unified Energy Dynamics (UED)**. This refined understanding connects science, philosophy, and metaphysics, reshaping how we view existence itself.

1. Nature of Energy

Traditional Approach: Energy is a measurable physical quantity associated with motion, heat, or potential within isolated systems, studied through frameworks like classical mechanics, thermodynamics, and quantum physics.

UED Approach: Energy is the **essence of all existence**, manifesting in dual states:

- **Stable Energy**: A timeless reservoir of latent potential.

- **Active Energy**: The dynamic driver of change. Energy acts as a **universal medium**, seamlessly connecting all phenomena across scales, bridging the tangible and intangible.

2. Conservation of Energy

Traditional Approach: Conservation laws state that energy cannot be created or destroyed but only transformed. These principles apply within localized systems, such as mechanical or thermodynamic processes.

UED Approach: Conservation operates **universally**: Active energy drives observable phenomena like motion, radiation, or chemical reactions. Stable energy absorbs disturbances to maintain universal equilibrium. This interplay ensures perpetual transformation within the infinite energy pool.

3. Transformation and Evolution

Traditional Approach: Energy transitions are seen as cause-and-effect processes occurring within defined boundaries, like chemical reactions or thermodynamic cycles. Evolution is attributed to stochastic or localized mechanisms.

UED Approach: Energy transitions act as the **engine of universal transformation**, creating a harmonious feedback loop between stability and dynamism. Cosmic evolution emerges as a synchronized interplay of energy, uniting quantum fluctuations and galactic formations into a coherent whole.

4. Entropy and Directionality

Traditional Approach: Entropy measures disorder, with systems naturally evolving toward greater entropy over time. The arrow of time is tied to entropy increase, emphasizing irreversibility.

UED Approach: Entropy reflects the **dispersal of active energy** and its reintegration into the stable energy pool. While entropy defines the direction of transformation, it also ensures continuity and equilibrium, preserving the universe's dynamic balance.

5. The Infinite Energy Pool

Traditional Approach: Energy interactions are studied as localized phenomena within defined systems, with no overarching concept of a universal energy medium.

UED Approach: The **infinite energy pool** is the unifying medium sustaining all existence. It encompasses both stable and active energy, providing the foundation for all physical and metaphysical phenomena. Acting as both source and sink, it governs energy transitions at all scales.

6. Consciousness and Energy

Traditional Approach: Consciousness is often viewed as a byproduct of brain activity or quantum interactions, studied in isolation within neuroscience, psychology, or quantum theories.

UED Approach: Consciousness is **energy in transit**, linking stable and active energy states: Human consciousness uniquely interacts with the energy medium through intellect and perception. Matter exhibits **quantum-level consciousness**, manifesting through its dynamic energy interactions.

7. Micro and Macro Unity

Traditional Approach: Separate frameworks govern quantum (microscopic) and relativistic (macroscopic) phenomena, and their unification remains an open challenge in modern physics.

UED Approach: Unified principles govern both micro and macro scales, with **energy dynamics** providing cohesion: From wave-particle duality to galactic structures, active and stable energy interactions shape all phenomena, revealing the interconnectedness of existence.

8. The First Disturbance

Traditional Approach: The Big Bang is posited as the singular event initiating the universe, without addressing the origin of energy or matter before it. Quantum superposition and consciousness are not integrated into this narrative.

UED Approach: The **First Disturbance** replaces the Big Bang, envisioning it as a **continual quantum superposition event**: This event initiated the interaction between stable and active energy.

It gave rise to photons, matter, and consciousness, marking the beginning of universal creation and sustained cosmic dynamics.

"The Transition"

The Chapter 3, **"The Transition,"** enhances Part I by examining exotic states of matter, such as **Bose-Einstein Condensates (BECs)** and **Fermionic Condensates**:

✓ These states, emerging at near-absolute zero, embody UED principles by transitioning from chaotic dynamics to quantum coherence.

✓ They offer insights into how stable and active energy interplay shapes matter and provides a glimpse into the early universe's conditions.

✓ Their study bridges theoretical physics and practical applications, from quantum computing to sustainable technologies, highlighting the tangible relevance of UED.

Thus, Part I redefines the narrative of existence through **Unified Energy Dynamics (UED)**, offering a cohesive framework that integrates science, philosophy, and metaphysics. It shifts the focus from fragmented, domain-specific views to a unified understanding of reality, where energy—manifesting in stable and active forms—acts as the cornerstone of all creation, transformation, and dissolution.

The inclusion of the **First Disturbance** enriches this framework, providing a profound perspective on the origin of the universe and the perpetual processes that sustain it. The addition of **The Transition** expands the scope, showcasing how exotic states of matter bridge quantum behavior and universal stability, further solidifying UED's relevance in connecting the tangible and intangible realms of existence.

"Do You Know?"

1. HOW DO STARS LIVE and Die?

Traditional Perspective: Stars form through the gravitational collapse of gas clouds, sustain themselves via nuclear fusion, and end as supernovae, neutron stars, or black holes.

UED Perspective: Stars emerge as disturbances in the stable energy medium convert into active energy, condensing matter into luminous celestial bodies. Their radiance reflects active energy interacting with the stable reservoir, creating the light and heat we observe. In their final stages, stars release active energy back into the infinite energy pool, completing the cosmic cycle of creation and dissolution.

2. Can Inanimate Matter Have Consciousness?

Traditional Perspective: Inanimate matter is considered devoid of consciousness, while biological frameworks study consciousness in living organisms.

UED Perspective: All matter interacts with the stable energy medium at a quantum level, exhibiting a subtle form of consciousness.Human consciousness is heightened due to the mind's functional manifestation, bridging intellect, perception, and active energy.This interaction reveals that existence is fundamentally tied to energy dynamics, uniting all entities in a shared energetic fabric.

3. Why Does Time Only Move Forward?

Traditional Perspective: Time flows in one direction, linked to entropy and irreversible processes.

UED Perspective: Time arises from human observation of energy transitions and changes in physical systems.Entropy, or energy dispersal, creates the illusion of a linear "arrow of time."Within the energy medium, time is non-existent; it is a human perception tied to the motion and transformations of active energy.

4. What Happens in a Black Hole?

Traditional Perspective: Black holes are regions where gravity is so intense that nothing, not even light, can escape.

UED Perspective: Black holes are zones where matter dissolves into stable energy, releasing active energy in the form of radiation. Hawking radiation may represent energy transitioning back to the stable energy pool. This process exemplifies the cycle of dissolution, as matter returns to the infinite energy medium, maintaining cosmic equilibrium.

5. Is Perpetual Motion Possible?

Traditional Perspective: Perpetual motion machines are deemed impossible due to energy losses and entropy.

UED Perspective: While isolated perpetual motion systems may not exist, the universe operates as a perpetually dynamic system. The continuous interplay of active and stable energy ensures a ceaseless flow, sustaining the cosmos's dynamic harmony.

6. What Explains Wave-Particle Duality?

Traditional Perspective: Particles exhibit wave and particle behaviors depending on observation, a cornerstone of quantum mechanics.

UED Perspective: Wave-particle duality reflects the interaction between active energy and the stable energy medium. Particles behave as waves during probabilistic transitions and as localized entities during interactions with matter. This duality reveals the dynamic and interconnected nature of the universal energy pool.

7. How Does the Energy Pool Shape Conscious Awareness?

Traditional Perspective: The universe is viewed as a vast, mostly empty space interspersed with matter and energy fields.

UED Perspective: The infinite energy pool pervades all existence, binding the universe into a unified whole. Human consciousness uniquely interacts with this medium, perceiving both tangible and intangible dimensions of reality. This connection enables humanity to explore the universe beyond physical limits, reflecting its special relationship with the energy pool.

8. Why Is Everything in the Universe Moving?

Traditional Perspective: Motion is attributed to initial conditions from the Big Bang, gravitational interactions, and conservation of momentum.

UED Perspective: Motion emerges from the inherent dynamic interplay between stable and active energy in the universal medium. Active energy drives localized disturbances, initiating motion at all scales—from subatomic particles to galaxies. Stable energy provides the equilibrium needed to channel this motion, ensuring coherence and continuity in the cosmic dance. This perpetual motion embodies the ongoing transformation and redistribution of energy, highlighting the universe's dynamic nature.

Conclusion

This section bridges traditional and UED perspectives, inviting readers to rethink fundamental questions about existence. By exploring the interconnected principles of energy dynamics, UED offers a cohesive understanding that transcends conventional boundaries, uniting science, philosophy, and metaphysics.

Part II: Time, Space, and the Cosmic Interplay

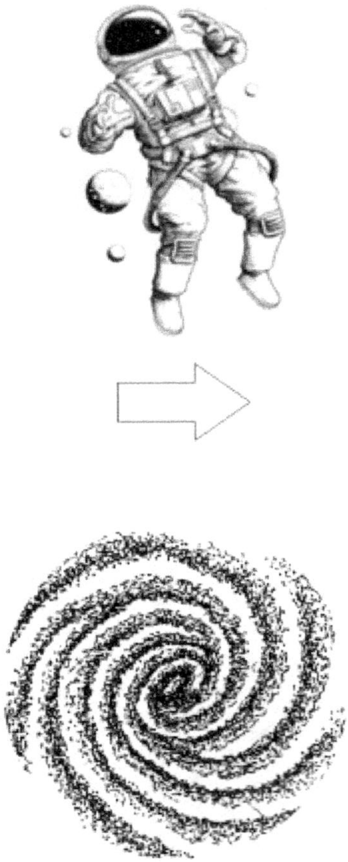

Chapter 6: Time: Human Perception in the Energy Medium

As we delve deeper into the essence of the universe, we confront a concept both familiar and enigmatic—time. In the previous chapters, we established energy as the cornerstone of existence, a universal medium that governs creation, transformation, and dissolution. Now, we turn to time, a phenomenon intrinsically linked to our observation of energy transitions yet profoundly misunderstood.

For humans, time serves as the framework within which events unfold—a linear progression marking the past, present, and future. However, within the infinite energy medium, time loses its traditional meaning, becoming an emergent property rather than a fundamental reality. By examining time through the lens of Unified Energy Dynamics (UED), we uncover its localized nature, its irrelevance within the infinite energy pool, and the speculative allure of concepts like time travel and multiverses. Time, as we perceive it, may be less a universal truth and more a reflection of consciousness interacting with the ever-dynamic dance of energy.

6.1 What Is Time? A Localized Phenomenon

Time, as we commonly perceive it, is a framework for organizing change—a human construct born out of the need to measure and understand the sequences of events. While clocks, calendars, and timelines create the illusion of a universal flow, time is, in reality, a localized phenomenon tied to the transitions and interactions of energy within physical systems.

Time as a Construct of Change and Observation

At its core, time emerges from the observation of change. The motion of celestial bodies, the oscillation of atomic particles, and even the beating of a human heart are all markers of energy in action. These markers provide a rhythm, a sequence that our minds interpret as the passage of time. Without these energy-driven changes, the concept of time would cease to hold meaning.

Human consciousness plays a pivotal role in this perception. By observing energy transitions—whether the rise and fall of the sun or the decay of radioactive elements—we create a narrative that defines past, present, and future. This narrative, however, is not an absolute truth but a localized interpretation unique to beings capable of conscious thought.

Energy Transitions and the Birth of Time

Time does not exist independently of energy; it is an emergent property of energy transitions within physical systems. When stable energy interacts with active energy, motion and transformation occur. These processes, whether as simple as the swing of a pendulum or as complex as the lifecycle of a star, generate measurable changes that form the basis of time.

On Earth, time is tied to cycles—day and night, the changing seasons, and the progression of life. These cycles are governed by energy dynamics, such as the Earth's rotation, the gravitational interplay between the Earth and the Moon, and the energy radiated by the Sun. At the quantum level, time arises from probabilistic interactions, such as the movement of particles within atomic systems.

Localized Nature of Time

Time is not universal; it is relative and context-dependent. Einstein's theory of relativity demonstrated that time dilates or contracts based on the observer's velocity and gravitational environment. This relativity underscores that time is not a fundamental aspect of the universe but a phenomenon deeply tied to localized energy conditions.

In the infinite energy medium proposed by Unified Energy Dynamics, time becomes irrelevant. The stable reservoir of energy, infinite and unchanging, does not experience time as humans understand it. Only in the realm of active energy—where motion, transformation, and interactions occur—does time emerge as a localized construct, a byproduct of the dynamic interplay between stability and motion.

Time, then, is not an absolute constant but a subjective lens through which we interpret the universe's unceasing dance of energy. It is a localized phenomenon, meaningful within physical systems yet irrelevant in the vast and timeless energy pool that binds the cosmos.

6.2 Time and the Infinite Energy Pool

In the vast expanse of the infinite energy pool, time as we perceive it ceases to hold relevance. Unlike the localized phenomena we associate with clocks, cycles, or decay, the stable energy reservoir exists beyond the confines of temporal progression. It is timeless, unchanging, and infinite—a foundational medium that transcends the human construct of past, present, and future.

The Irrelevance of Time in the Infinite Energy Medium

Time emerges from the observation of change, yet the stable energy pool represents constancy and equilibrium. Within this reservoir, energy does not move, transform, or dissipate—it exists as a latent potential, a silent foundation upon which all dynamic interactions occur. In this realm, there is no "before" or "after," only an eternal state of presence.

Active energy, by contrast, is the source of motion and transformation, generating the observable changes that give rise to time. However, these changes are confined to localized systems, whether the orbit of a planet, the decay of an atom, or the life of a star. The infinite energy pool remains untouched by these transient disturbances, maintaining its timeless nature.

Energy Dynamics Beyond Temporal Constraints

In the infinite energy pool, energy transitions do not follow a linear timeline; they exist as cyclical and interconnected processes. Disturbances in the stable energy medium give rise to active energy, which manifests as motion, heat, light, and other observable phenomena. When these disturbances dissipate, active energy returns to the stable reservoir, completing a cycle of creation and dissolution.

This cyclical nature defies the forward-moving arrow of time perceived in localized systems. Instead, it represents a dynamic equilibrium, where energy is neither created nor destroyed but perpetually redistributed. In this context, time is not a governing principle but an emergent property tied to the localized interactions of energy.

Timelessness as a Universal Principle

The absence of time in the infinite energy pool reveals a deeper truth about the universe: time is not a universal constant but a relative experience tied to specific conditions. In the timeless expanse of the energy medium, all phenomena—past, present, and future—exist as interconnected manifestations of energy's eternal dance.

This understanding shifts the narrative of time from a linear progression to a localized construct, applicable only within dynamic systems. It invites us to reimagine the universe as a timeless and interconnected whole, where energy dynamics govern all existence, free from the constraints of temporal progression.

6.3 The Illusion of Time Travel

Time travel has long captured the imagination of scientists and storytellers alike, envisioning journeys into the past or leaps into the future. However, while compelling in theory and fiction, the concept of time travel becomes untenable when examined through the principles of energy dynamics, entropy, and conservation. The very framework that sustains the universe undermines the feasibility of traversing time as if it were a navigable dimension.

The Speculative Nature of Time Travel

The idea of time travel stems from the perception of time as a linear, absolute continuum—one that can be reversed or accelerated. In this view, traveling backward could allow revisiting past events, while forward travel could provide glimpses into the future. However, such interpretations rely on a simplistic understanding of time and ignore the intricate interplay of energy transitions and entropy.

Time, as established in the UED framework, is not an independent dimension but a localized phenomenon arising from energy transitions and human perception of change. Consequently, the notion of "traveling" through time becomes irrelevant in the absence of a universal timeline.

Entropy: The Irreversible Arrow

A fundamental obstacle to time travel lies in the principle of entropy—the measure of energy dispersal within a system. Entropy always increases in a closed system, creating the "arrow of time" that defines the irreversibility of processes. Reversing time would require reversing entropy, effectively undoing the natural progression of energy dispersal.

For example: To travel backward in time, every interaction at the atomic and cosmic level would need to reverse direction—a feat that contradicts the inherent tendency of systems to evolve toward greater disorder. Similarly, Forward time travel, as imagined in science fiction, implies jumping beyond the natural flow of energy transitions, disrupting the equilibrium of the energy medium.

Entropy not only resists such reversals but also ensures that the universe remains in a state of dynamic harmony, governed by the constant interplay of stability and motion.

Energy Conservation: A Barrier to Time Travel

The principle of energy conservation further challenges the feasibility of time travel. Energy transitions occur within a closed system, with active energy converting into various forms while stable energy absorbs disturbances. Any attempt to travel through time would require manipulating these transitions on an unprecedented scale, violating the law of conservation.

Consider backward time travel:

✓ To reconstruct the past, active energy would need to revert to its earlier states, undoing transformations that have already contributed to the universe's evolution.

✓ This would demand an infinite reservoir of energy to reverse every interaction, an impossibility within the framework of conservation.

Similarly, forward travel would involve bypassing natural energy transitions, disrupting the equilibrium of the universal energy pool and destabilizing localized systems.

The Reality Behind the Illusion

The allure of time travel lies in its imaginative potential, but the universe operates on principles that render it an illusion. Entropy and conservation do not merely constrain time travel—they reaffirm the interconnected nature of existence, where every moment is shaped by the continuous flow of energy.

In the UED perspective, time is a construct tied to observation and energy dynamics, not a dimension to be navigated. Instead of focusing on speculative travel through time, this understanding invites us to explore the timeless energy pool, where creation, transformation, and dissolution unfold in an eternal, interconnected dance.

6.4 Multiverses: A Speculative Escape

The concept of multiverses—alternate realities existing alongside our own—has captivated imaginations and provided fertile ground for scientific theories and speculative fiction. Often tied to time-based constructs, multiverse theories suggest the existence of infinite, branching timelines where every possibility manifests in a parallel universe. While this idea offers an enticing escape from the constraints of a singular reality, it lacks grounding when viewed through the lens of Unified Energy Dynamics (UED).

Multiverse Theories and Time-Based Constructs

Multiverse theories often emerge from interpretations of quantum mechanics or speculative cosmology:

Quantum Interpretations: In the "many-worlds" view, every quantum decision or interaction leads to the creation of a new universe, forming an ever-branching tree of alternate realities.

Cosmic Speculations: In some cosmological models, the universe's initial conditions give rise to multiple, independent universes, each with its own set of physical laws.

Both interpretations hinge on the idea of time as a linear, branching phenomenon, where each decision or event spawns divergent timelines. These constructs imply that time is not only navigable but also generative, creating new universes at every juncture.

UED's Perspective on a Singular, Infinite Medium

From the standpoint of UED, the notion of multiverses conflicts with the foundational principles of an infinite, interconnected energy medium:

Energy Unification: The universe, as described by UED, is a singular, cohesive system where all phenomena emerge from and return to the infinite energy pool. This medium, encompassing both stable and active energy, negates the need for separate realities or branching timelines.

No Alternate Timelines: Time, in the UED framework, is not an absolute dimension but a localized phenomenon tied to energy transitions. Without an independent or universal timeline, the concept of alternate timelines becomes irrelevant, as time itself is a perception rather than a fundamental property of the cosmos.

Infinite but Singular: The universe's infinity lies not in the multiplicity of worlds but in the boundless nature of its energy medium. This singular infinity sustains all existence within one coherent framework, making alternate universes unnecessary for explaining cosmic complexity.

The Illusion of Separation

Multiverse theories may provide imaginative narratives, but they overlook the interconnectedness of existence within a unified energy pool. By positing separate realities, these theories fragment the cosmos, disregarding the inherent harmony of energy dynamics:

Observable phenomena across all scales—from quantum interactions to galactic formations—arise within the same energy medium, governed by universal principles.

The stability and continuity of the energy pool ensure that creation, transformation, and dissolution occur within a singular reality.

Embracing the Singularity

In the UED perspective, the universe's complexity and diversity do not require multiple realities or alternate timelines. Instead, they are manifestations of the infinite potential within a singular energy medium. This approach shifts the focus from speculative escapes to understanding the profound interconnectedness of all things, where every fluctuation and transformation is a part of one unified existence.

By embracing this singular framework, we move closer to uncovering the universe's true nature—an infinite, cohesive system where energy serves as the foundation of all creation and harmony.

6.5 Time as Perceived by Conscious Entities

Time, as experienced by humans and other conscious beings, is not an inherent property of the universe but a perception shaped by the mind's ability to observe and interpret energy transitions. This perception creates the illusion of linearity, allowing conscious entities to navigate the world, make decisions, and organize experiences within the constructs of "past," "present," and "future." Unified Energy Dynamics (UED) reveals that time, far from being universal, is a localized phenomenon tied to consciousness and its interaction with the energy medium.

Human Consciousness and the Observation of Time

Consciousness acts as the lens through which energy dynamics are observed, giving rise to the perception of time:

Observation of Change: Time is perceived through the observation of changes in energy states, such as the motion of celestial bodies, the ticking of a clock, or the growth of living organisms. These changes provide markers that allow conscious entities to distinguish "before" from "after," forming the foundation of the temporal experience.

Sequential Awareness: The human mind processes events in a sequential manner, creating the impression of a linear flow of time. This sequential awareness enables planning, memory formation, and a sense of continuity, vital for survival and cognitive function.

The Illusion of Linearity in a Timeless Framework

While time appears linear to conscious beings, this perception arises from the limitations of observation and the mind's structuring of information:

Localized Phenomenon: Time exists only within the realm of physical systems, where energy transitions create observable changes. Beyond these localized interactions, such as within the infinite energy pool, time loses its relevance, as stability transcends the need for temporal markers.

Construct of the Mind: The mind organizes sensory input and experiences into a chronological order, creating the appearance of a "timeline." This construct allows for the practical navigation of reality but does not reflect the true nature of the cosmos, which operates outside the constraints of time.

Timeless Reality: In the framework of UED, the universe is inherently timeless, with all phenomena occurring within the interplay of stable and active energy. The illusion of time arises only when conscious entities interact with and interpret these dynamics, anchoring their experiences in a perceived temporal flow.

Consciousness: The Creator of Time

The relationship between consciousness and time reveals the profound role of conscious entities in shaping their understanding of reality:

Energy in Transit: Consciousness, as energy in transit between stable and active states, bridges the timeless energy medium and the dynamic processes that create the perception of time. This transit allows conscious beings to engage with the universe on both tangible and intangible levels.

Temporal Constructs: By interpreting energy transitions, consciousness creates temporal constructs that guide decision-making and interaction with the physical world. These constructs, while practical, remain subjective and limited to the conscious experience, highlighting their illusory nature in a timeless framework.

A New Perspective on Time

Understanding time as a construct shaped by consciousness shifts our perception of reality:

✓ It challenges the notion of time as a fundamental property of the universe, redefining it as a localized experience tied to energy transitions.

✓ It underscores the unique role of conscious entities in observing, interpreting, and interacting with the cosmos, where timeless energy dynamics form the true foundation of existence.

Through this lens, time becomes a tool of perception rather than an absolute truth, illuminating the intricate relationship between consciousness and the universal energy medium.

6.6 The Arrow of Time: A Product of Entropy

The unidirectional flow of time, often referred to as the "arrow of time," is deeply tied to the concept of entropy. In the framework of Unified Energy Dynamics (UED), this arrow is not an inherent property of the universe but a manifestation of energy dispersal within physical systems. As entropy increases, the perception of time as a one-way progression emerges, shaped by the irreversible nature of energy transitions and the tendency of systems to evolve toward greater states of disorder.

Entropy and the Directionality of Time

Entropy, a measure of energy dispersal and system disorder, provides the foundation for time's apparent unidirectional flow:

Irreversible Processes: Energy transitions, such as the flow of heat from hot to cold or the dissipation of motion into thermal energy, are inherently irreversible. This irreversibility creates a sense of progression, where systems move from ordered states to less ordered configurations, shaping the perception of "before" and "after."

Energy Dispersal as the Driver: The dispersal of active energy into the stable reservoir reflects the natural tendency of systems to reach equilibrium. This process underpins the observable flow of events, where energy spreads and transforms, guiding the evolution of physical systems.

Time as a Reflection of Energy Dynamics

The arrow of time emerges from the dynamic interplay between energy transitions and entropy:

Entropy as a Marker: Increasing entropy serves as a marker for the passage of time, allowing conscious entities to perceive change and sequence. **For example:** In the lifecycle of a star, nuclear fusion reduces order, converting mass into energy and increasing entropy. Similarly, in daily life, processes like combustion, growth, and decay showcase energy dispersal, reinforcing the perception of time's forward march.

Physical Reality Anchored in Energy: Time exists as a reflection of energy's behavior within the physical realm, where transitions and transformations dominate. In the infinite energy pool, where stability transcends transitions, time loses relevance, highlighting its localized nature.

Irreversibility and Perception: The inability to reverse entropy-driven processes creates the illusion of time as a one-way flow. This unidirectionality, however, is a construct of observation rather than an absolute feature of the cosmos.

The Arrow of Time and Human Perception

Human consciousness plays a pivotal role in interpreting the arrow of time, tying it to observed changes and energy transitions:

Sequential Awareness: The mind organizes sensory input into a chronological order, creating the impression of past, present, and future. This sequential awareness aligns with entropy's increase, reinforcing the idea of time's linearity.

Entropy as a Cognitive Anchor: The progression of entropy provides a consistent framework for interpreting events, guiding decision-making and the organization of experience. Without entropy-driven transitions, the perception of time would lack a tangible reference point, dissolving into the timeless energy medium.

A Dynamic Reflection of Reality

The arrow of time, as understood through UED, is not a fundamental property but a dynamic reflection of energy dispersal:

In Physical Systems: It highlights the interplay of stability and motion, where entropy governs the evolution of systems, from the quantum scale to the cosmic.

In Human Experience: It serves as a tool for navigating reality, providing a framework for understanding change and continuity.

By recognizing the arrow of time as a product of entropy, we uncover its true nature—a localized phenomenon that emerges from energy's eternal dance. This perspective reshapes our understanding of time, grounding it in the universal principles of energy dynamics.

6.7 The Role of Time in Cosmic Harmony

Time, as a construct tied to human perception, plays a vital role in structuring experience and understanding. It allows conscious entities to organize observations, interpret change, and navigate the complexities of physical reality. However, in the grander scheme of the cosmos, time is not a governing principle but a derivative of energy transitions and entropy. Through this dual lens—its utility for humans and its irrelevance in the infinite energy pool—we explore the nuanced role of time in maintaining cosmic harmony.

Time as a Tool for Human Understanding

For conscious entities, time provides a framework to interpret the universe's dynamic processes:

Organizing Experience: Time enables humans to categorize events into past, present, and future, creating a sense of continuity and causality. This perception allows for the development of knowledge, memory, and foresight, which are essential for survival and growth.

Guiding Interpretation: The observation of energy transitions, such as the rising and setting of the sun or the changing seasons, gives time a tangible presence in daily life. These cycles offer reference points, helping humanity comprehend the universe's rhythms and align actions with natural processes.

Facilitating Progress: The linear concept of time underpins scientific inquiry and technological advancement, providing a framework for measuring and understanding phenomena. Time's perceived flow helps humans track progress, plan for the future, and reflect on the past.

Timeless Principles Governing the Cosmos

While time serves as a practical tool for human understanding, the cosmos itself operates beyond temporal constraints:

Energy as the Fundamental Driver: The universe is governed by energy dynamics—transitions between stable and active states, interactions within the infinite energy pool, and the dispersal of energy through entropy. These processes occur independently of time, guided by timeless principles that ensure cosmic equilibrium.

Cosmic Cycles and Harmony: The formation and dissolution of stars, the propagation of gravitational waves, and the expansion of galaxies follow patterns dictated by energy interactions, not temporal progression. These cycles reveal a universe in perpetual motion, where the balance between stability and dynamism sustains harmony.

Transcending Localized Perceptions: Time, as perceived by humans, is a localized phenomenon arising from the observation of change. In the infinite energy medium, where stability underpins all existence, time loses its significance, highlighting the timeless nature of universal laws.

The Interplay Between Time and Timelessness

The duality of time's relevance illustrates its role as both a practical tool and a conceptual limitation:

For Conscious Entities: Time bridges the gap between human perception and the infinite complexity of the cosmos, offering a way to navigate reality.

In the Cosmic Context: The universe's harmony is rooted in energy principles that transcend the need for a temporal framework, underscoring the inherent unity of existence.

Time, while indispensable for structuring human experience, is not a fundamental feature of the universe. Its localized perception arises from the observation of energy transitions, providing a lens through which humans engage with the cosmos. Yet, the timeless principles of energy dynamics govern the universe's behavior, ensuring its balance and harmony. By understanding time's dual role, we reconcile its utility in daily life with its irrelevance in the infinite energy pool, bridging human experience with the cosmic reality.

Summary: Time: Human Perception in the Energy Medium

This chapter explores the concept of time through the lens of **Unified Energy Dynamics (UED)**, redefining it as an emergent and localized phenomenon tied to energy transitions rather than a universal constant.

1. Time as a Localized Phenomenon

Time is not an absolute, universal entity; instead, it is a **human construct** created to organize change and observe the sequences of events.

Energy Transitions as the Basis: Time arises from the motion and transformation of energy, such as the oscillation of particles or the decay of elements. Without these transitions, time would cease to hold meaning.

2. Localized Nature of Time

Time is **relative and context-dependent**, as demonstrated by Einstein's relativity, where gravitational fields and velocity influence time dilation and contraction.

In the **infinite energy pool** of UED, time becomes irrelevant because the stable energy medium exists beyond transitions or entropy changes.

3. Human Perception of Time

Human consciousness plays a pivotal role in shaping the perception of time: Observing changes like the motion of celestial bodies or the ticking of clocks creates a narrative of past, present, and future. This perception is a **localized interpretation** unique to beings capable of conscious observation.

4. Time in the Infinite Energy Medium

The infinite energy pool is **timeless**: Stable energy, as a constant reservoir, transcends time and exists in equilibrium. Only within localized systems of active energy do transformations generate the flow of time.

5. The Arrow of Time

Time's unidirectional flow (the arrow of time) is tied to **entropy**, where systems evolve toward equilibrium.

This progression of entropy creates the perception of a forward-moving timeline but is a reflection of localized dynamics, not a universal truth.

6. Time Travel and its Feasibility

Time travel, as traditionally imagined, becomes irrelevant in the UED framework: Reversing entropy or disrupting energy transitions contradicts the natural flow of the universe. The idea of a "universal timeline" is replaced by localized perceptions of time within systems.

7. Dual Role of Time

For Humans: Time provides a framework for organizing experiences, guiding decision-making, and interpreting reality.

In the Cosmic Context: Time is a localized construct, unnecessary for the timeless dynamics of the infinite energy pool.

Time, as understood through UED, is not a fundamental property of the universe but a reflection of localized energy interactions and human perception. While essential for structuring human experience, time becomes irrelevant within the timeless energy medium that governs the cosmos.

Chapter 7: Space: The Infinite Stage of Energy

As we journey further into understanding the universe through Unified Energy Dynamics (UED), we arrive at space—the infinite stage upon which energy and matter interact. Space is not an empty void but the steady framework that supports the cosmos. It is the boundless medium that provides structure and coherence to all phenomena, from the motion of particles to the formation of galaxies.

In previous chapters, we explored energy as the universal medium and time as a localized perception arising from energy transitions. Now, we turn our focus to space, which serves as the constant backdrop for these dynamics. Space allows for the expression of energy in its stable and active states, giving rise to the physical reality we perceive. By redefining space through the lens of UED, we uncover its true nature—not as a passive container but as an active participant in the universe's eternal dance of creation and transformation.

7.1 What Is Space? The Framework of Existence

Space is often imagined as a vast, empty void—a limitless expanse in which objects exist and events unfold. However, from the perspective of Unified Energy Dynamics (UED), space is far more profound. It is not merely a passive backdrop but an active and stable framework that supports the universe's existence. Space provides the structure within which energy transitions occur, matter interacts, and phenomena manifest.

At its core, space embodies stability. It acts as the infinite, unchanging foundation that allows the dynamic interplay of energy and matter to unfold. Without this framework, the transformations and motions driven by active energy would lack coherence, and the universe would descend into chaos. Space ensures that energy's dual states—stable and active—can interact harmoniously, maintaining balance and sustaining the intricate systems we observe.

In this sense, space is both the stage and the medium. It does not merely accommodate existence; it actively shapes it by providing the conditions necessary for creation, transformation, and dissolution. From the quantum scale to the cosmic, space is the steady backdrop that binds the universe into a cohesive whole.

7.2 Mass as Disturbances in the Energy Medium

In the framework of Unified Energy Dynamics (UED), mass is not a fundamental entity but rather a manifestation of localized disturbances in the infinite energy medium. Stable energy, the silent reservoir underpinning the universe, remains undisturbed in its natural state. However, when disruptions occur—whether due to quantum fluctuations or larger-scale cosmic events—these disturbances condense into mass, giving rise to the physical structures we observe in the universe.

Mass represents active energy that has become concentrated and localized. This concentration introduces curvature to space, shaping the interactions between energy and matter. For example:

Formation of Celestial Bodies: Stars, planets, and galaxies emerge as regions of condensed active energy, coalescing from the stable medium during disturbances. These masses create gravitational effects, further organizing matter into cosmic structures.

Quantum-Level Manifestations: At the smallest scales, particles like electrons and quarks also arise as disruptions within the stable energy pool, maintaining dynamic equilibrium through their interactions.

Localized disturbances, while appearing as independent entities, are deeply connected to the surrounding stable energy medium. This connection ensures that mass is not static or isolated but part of a larger interplay that sustains the universe's harmony.

The presence of mass influences the energy medium, creating ripples and interactions that drive cosmic evolution. These disturbances form the scaffolding upon which the observable universe is built, from the smallest particles to the largest galactic formations. Yet, just as mass emerges from disruptions in the stable energy pool, it also returns to this pool—dissolving back into stability through processes like black hole collapse or the dispersal of matter during cosmic events.

Through this perspective, mass is reimagined not as a fundamental constant but as a dynamic participant in the universe's eternal dance of creation, transformation, and dissolution. It is the visible expression of energy's interplay with space, revealing the profound interconnectedness of all existence.

7.3 Rethinking Wormholes: Voids in the Energy Medium

Wormholes, often romanticized as shortcuts through spacetime, are among the most captivating concepts in modern cosmology. Traditional interpretations, rooted in Einstein's General Relativity, envision wormholes as hypothetical passages connecting distant regions of the universe. These constructs have fueled imaginative scenarios of interstellar travel and even time manipulation. However, through the lens of Unified Energy Dynamics (UED), wormholes take on a radically different character—one that strips away their mystique and redefines them as voids within the energy medium.

Wormholes as Energy-Devoid Regions

In the UED framework, the universe is bound together by an infinite energy medium, composed of stable energy that provides coherence and active energy that drives change. Wormholes, if they exist, represent anomalies where the energy medium is absent or significantly disrupted.

Absence of Energy: Wormholes can be viewed as voids—regions devoid of both stable and active energy. Unlike black holes, which act as reservoirs where matter and energy are absorbed and eventually reintegrated into the stable pool, wormholes lack the energy infrastructure necessary to sustain their existence or interact meaningfully with surrounding phenomena.

Breakdown of Continuity: The absence of energy in these regions disrupts the fundamental interconnectedness of the cosmos, rendering wormholes incompatible with the energy dynamics that govern universal harmony.

Why Wormholes Are Not Shortcuts Through Spacetime

The allure of wormholes lies in their portrayal as shortcuts, enabling instantaneous travel across vast distances or even between timelines. However, such interpretations falter when scrutinized through the principles of energy conservation, stability, and entropy:

Energy Infrastructure Is Missing: For a wormhole to serve as a passage, it would require an underlying energy framework to support the transit of matter or energy. The absence of stable energy within these voids precludes the possibility of a coherent structure capable of maintaining such a passage.

Incompatibility with Entropy: Wormholes contradict the second law of thermodynamics, which dictates an increase in entropy and the irreversibility of energy transitions. The concept of traversing a wormhole defies these principles, as it implies a form of energy manipulation that violates natural laws.

Disruption of Universal Continuity: UED emphasizes the universe as a cohesive, interconnected system where energy transitions occur in harmony. Wormholes, as isolated voids, disrupt this continuity and lack the dynamic interplay required to integrate with the universal framework.

A New Perspective on Wormholes

Instead of viewing wormholes as pathways, UED redefines them as localized disruptions or voids in the energy medium—regions where the flow of stable and active energy is either nullified or heavily distorted. These voids do not connect regions of spacetime but instead represent anomalies that highlight the limits of our understanding of energy's behavior under extreme conditions.

Through this perspective, wormholes are no longer gateways to alternate realities or distant corners of the universe. They are instead conceptual tools that challenge us to rethink the fabric of existence and the role of energy in maintaining universal harmony. Rather than shortcuts, they stand as reminders of the infinite complexity of the cosmos and the limitations of speculative constructs when divorced from the foundational principles of energy dynamics.

7.4 The Infinite Nature of Space and Energy

Space, often perceived as a vast and empty expanse, is far more than a void separating celestial objects. In the Unified Energy Dynamics (UED) framework, space emerges as the infinite stage where the intricate dance of energy and matter unfolds. It is not just a backdrop but an active participant in shaping the cosmos, integrating the micro and macro realms into a unified, harmonious whole.

Space as the Boundless Stage for Energy Transformations

Space serves as the steady, infinite canvas upon which energy transitions and interactions occur. It is the medium that accommodates the interplay between stable and active energy, providing a framework for the creation, transformation, and dissolution of matter.

Unconfined by Boundaries: Unlike localized phenomena, space extends infinitely, offering an unrestricted environment for energy dynamics to manifest. This boundlessness ensures that the processes of motion, radiation, and interaction occur seamlessly, connecting all points in the universe.

Facilitator of Transformation: Within space, active energy drives observable phenomena—whether it be the formation of galaxies or the motion of subatomic particles—while stable energy maintains equilibrium, ensuring the coherence of the cosmic fabric.

A Dynamic Partner: Far from being static, space interacts with energy, responding to its transitions. Gravitational waves, for instance, ripple through spacetime, demonstrating how space itself participates in the propagation of energy.

Integrating the Micro and Macro Realms

One of the most profound aspects of space is its ability to bridge the smallest quantum interactions with the grandest cosmic structures. It is the unifying medium where the same fundamental principles govern both scales, revealing a deeply interconnected reality.

At the Micro Scale: Space provides the environment for quantum phenomena such as wave-particle duality and entanglement. These behaviors, though seemingly localized, are shaped by the infinite energy medium, showcasing the seamless integration of energy and space.

At the Macro Scale: Space is the arena where galaxies form, stars evolve, and black holes exert their influence. These cosmic events are not isolated but are expressions of the same energy dynamics observed at the quantum level, magnified on a universal scale.

A Unified Whole: The infinite nature of space ensures that no phenomenon is disconnected. From the flickering of quarks to the gravitational dance of celestial bodies, all actions occur within the same boundless medium, tied together by the laws of energy dynamics.

The Infinite Cosmos: A Reflection of Unity

Space's infinite nature reflects the universe's inherent unity, where every transformation, no matter the scale, contributes to the grand tapestry of existence. By accommodating the perpetual motion of energy and matter, space acts as the ultimate integrator, dissolving distinctions between the microscopic and the macroscopic.

In this light, space is not merely a container but the very essence of connection. It is where the infinite energy pool manifests its presence, shaping and sustaining the cosmos in a state of dynamic equilibrium. Through this understanding, we recognize space as the bridge that unites all realms of existence, weaving the universe into a single, cohesive whole.

7.5 Space, Matter, and the Dance of Creation

Space and matter are intrinsically connected in a perpetual interplay that defines the universe's structure and behavior. In the Unified Energy Dynamics (UED) framework, space is not merely a passive backdrop but an active participant in the processes of creation, transformation, and dissolution. The relationship between space and matter exemplifies the dynamic nature of energy transitions, where disturbances in the energy medium give rise to physical forms, and those forms, in turn, influence the fabric of space.

The Dynamic Relationship Between Space and Matter

Matter emerges as localized disturbances in the infinite energy medium, creating ripples in the otherwise stable pool of energy. This interaction is not one-sided; space actively responds to the presence of matter, shaping its behavior and defining its motion.

Space as the Stage: Space provides the structural framework that supports matter's existence. It accommodates energy transitions, allowing disturbances in the stable energy pool to coalesce into mass, forming particles, stars, and galaxies.

Matter as the Sculptor: Matter, in turn, modifies space through its interactions with energy. Gravitational forces, for example, arise from the interaction of mass with the stable energy medium, warping space and creating pathways for motion and interaction.

The Interdependent Dance: Neither space nor matter exists in isolation. Their dynamic relationship is akin to a dance, where space offers stability and structure while matter introduces motion and transformation.

How Space Shapes Energy Transitions

Space plays a pivotal role in channeling and sustaining energy transitions, ensuring that disturbances in the energy medium result in coherent and observable phenomena.

Energy Flow: The infinite expanse of space allows energy to propagate as waves, whether in the form of light, gravitational ripples, or quantum fields. These energy flows are guided by the structure of space, ensuring their reach across vast distances.

Formation of Structures: As energy transitions give rise to matter, space shapes its arrangement and motion. The clustering of galaxies, the orbital paths of celestial bodies, and the dynamics of molecular interactions are all influenced by space's steady framework.

How Matter Shapes Space

While space influences energy transitions and the formation of matter, matter, in turn, exerts a profound effect on space, demonstrating their mutual dependence.

Curvature of Space: Matter, as a concentrated disturbance in the energy medium, warps the fabric of space. This curvature determines how objects move, influencing everything from planetary orbits to the trajectories of light near massive celestial bodies.

Energy Redistribution: The presence of matter redistributes energy within space, creating zones of interaction where transformations and transitions occur. Black holes, for instance, represent extreme distortions in the energy medium, where matter dissolves into stable energy, altering the surrounding space.

The Perpetual Dance of Creation

The relationship between space and matter is not static; it is a continuous process of creation and interaction driven by energy dynamics.

Emergence and Dissolution: Matter emerges from disturbances in the stable energy pool, and as those disturbances dissipate, matter returns to the pool, completing the cycle of existence. Space, as the infinite stage, accommodates and sustains this cycle.

Cosmic Harmony: The interplay of space and matter ensures the universe's dynamic equilibrium, where transformation and stability coexist. This dance of creation highlights the interconnectedness of all phenomena, bound together by the principles of energy.

Through this lens, space and matter are not separate entities but partners in a cosmic ballet. Together, they manifest the principles of energy dynamics, shaping the universe into a cohesive and ever-evolving reality.

7.6 The Bending of Light and the Role of the Energy Medium

The bending of light around massive objects, popularly attributed to Einstein's theory of General Relativity, offers profound insights when viewed through the lens of Unified Energy Dynamics (UED). While classical physics describes this phenomenon as the warping of spacetime by gravity, UED provides a complementary perspective, highlighting the role of the universal energy medium in shaping this interaction.

Light's Path in the Energy Medium

At its core, light is a form of electromagnetic radiation—waves of energy that interact with the universal energy medium. As light traverses space, its path is not inherently curved; it moves in straight lines unless influenced by external factors. Massive celestial objects, such as stars and black holes, introduce localized disturbances in the stable energy pool, creating what is perceived as spacetime curvature. However, in UED, this curvature is better understood as the localized deformation of the energy medium.

Mass Disturbances and Energy Medium Dynamics: Massive objects disrupt the equilibrium of the energy medium, causing it to "bend" around them. Light, traveling through this deformed medium, naturally follows the altered trajectory. This bending is not due to a direct gravitational pull on photons (which are massless), but rather the change in the structure of the medium itself.

Uniform Energy Fields vs. Disturbed Regions: In uniform energy fields, light moves in a straight line, undisturbed by localized fluctuations. In contrast, regions near massive objects present an "energetic gradient," guiding light along a curved path. This process demonstrates the intricate relationship between energy medium dynamics and observable phenomena.

The Illusion of Gravitational Force on Light

Einstein's framework treats gravity as the curvature of spacetime caused by mass, which indirectly governs the trajectory of light. UED refines this idea by emphasizing that the energy medium—disturbed by mass—is the actual agent influencing light's path. Gravity, redefined as the interaction between mass and the stable energy medium, creates conditions for the bending of light:

Photon Interaction with the Medium: Photons, as packets of active energy, interact with the energy medium without relying on mass-based forces. Their trajectory is influenced by the "topography" of the medium, shaped by nearby masses. This interaction ensures that light bends in a manner consistent with gravitational lensing, as observed in astrophysical studies.

Cosmic Implications: The bending of light due to the energy medium's dynamics has profound implications for understanding cosmic distances and observations. Events like gravitational lensing, where massive objects magnify and distort the light of background galaxies, can now be viewed as evidence of the energy medium's role in mediating cosmic interactions.

The Stability of the Energy Medium in Light Propagation

The energy medium serves as the universal stage for light propagation, ensuring coherence and stability despite disturbances. This stability explains why light travels at a constant speed in a vacuum, unaffected by variations in mass or energy concentration, except in regions of extreme disturbance like black holes.

The Role of Energy Pool Stability: The energy medium, while adaptable to disturbances, maintains an overarching equilibrium that supports the consistent behavior of light across vast cosmic distances. The bending of light, therefore, is a localized phenomenon rather than a universal rule, tied to specific conditions of mass-energy interaction.

Why Light Fails in Black Holes: In extreme cases, such as black holes, the energy medium's disturbance reaches a critical point where photons cannot escape. Here, light's trajectory loops into the singularity, merging active energy back into the stable pool, as per UED principles.

Bridging Classical and UED Perspectives

While Einstein's General Relativity laid the foundation for understanding light bending through spacetime curvature, UED enriches this narrative by introducing the energy medium as the fundamental agent. This framework bridges the gap between observational evidence and theoretical constructs, offering a cohesive explanation of light's behavior in the universe.

Through this lens, the bending of light is no longer a mere consequence of gravity but a testament to the universal energy medium's dynamic nature. It reaffirms the interconnectedness of light, mass, and energy, showcasing how disturbances in the medium create phenomena that shape our understanding of the cosmos.

7.7 Conclusion: Space as Energy's Eternal Stage

Space is not merely an empty void; it is the eternal canvas upon which the universe paints its dynamic masterpiece. As the stable framework of existence, space enables energy to transform, matter to emerge, and the cosmos to evolve in a harmonious interplay of creation and dissolution. Far from being passive, it is an active participant, sustaining the intricate balance between stability and motion, the tangible and intangible.

Within the Unified Energy Dynamics (UED) framework, space takes on a transformative role as the infinite reservoir that integrates energy and matter into a unified whole. It is the medium where disturbances in the stable energy pool give birth to matter and motion, while maintaining the coherence that binds the micro and macro realms. Space nurtures every ripple in the energy medium, guides the emergence of celestial bodies, and orchestrates the cosmic dance of forces.

For humanity, space represents both an enigma and an opportunity—a vast frontier that challenges our limited perceptions and invites us to explore its infinite depths. While our understanding is shaped by time-bound observations and localized experiences, UED reveals space as a reality that transcends these constraints. It bridges the gap between human perception and the boundless cosmos, unveiling a universe that is not fragmented but profoundly interconnected.

Space, as energy's eternal stage, is the silent architect of cosmic harmony, the unseen foundation upon which existence rests. It binds the universe into a dynamic and ever-evolving unity, ensuring that every transformation operates within the timeless principles of energy. By embracing space's true nature, we unlock a deeper understanding of the universe—one that celebrates its infinite beauty, coherence, and order. It is through this lens that space becomes not just a setting, but the very essence of the cosmos' harmonious existence.

Chapter 8: Gravity and the Cosmic Tapestry

As we journey deeper into the understanding of the universe, we encounter a force that has captivated human thought for centuries: gravity. Traditionally described as the attraction between masses, gravity is often viewed as the architect of celestial order—binding planets, stars, and galaxies into structured harmony. However, through the lens of Unified Energy Dynamics (UED), gravity emerges as far more than a mass-based force. It is a manifestation of the interaction between mass and the infinite energy medium, a dynamic interplay that extends beyond the limitations of traditional frameworks.

In this chapter, we delve into gravity's role within the cosmic tapestry, exploring its nature as the weakest yet most pervasive force in the universe. By redefining gravity as an outcome of energy dynamics, we uncover its deeper connection to the equilibrium of the cosmos, bridging the micro and macro realms and shedding new light on phenomena like black holes and gravitational waves. Through this perspective, gravity is no longer just an attractive force but a cornerstone of the intricate dance that sustains universal harmony.

8.1 What Is Gravity? A New Perspective

Gravity, traditionally understood as the force of attraction between masses, has been the cornerstone of classical physics since Newton's time. However, Unified Energy Dynamics (UED) offers a revolutionary redefinition of gravity, moving beyond the constraints of mass-based attraction to reveal its deeper essence. In the UED framework, gravity emerges not as an isolated force but as a manifestation of the interaction between mass and the stable energy medium.

At its core, gravity is the result of disturbances in the stable energy pool caused by the presence of mass. These disturbances create localized interactions that manifest as the gravitational "pull" experienced between objects. Rather than being a fundamental force in isolation, gravity is a relational phenomenon—a ripple in the infinite energy medium generated by mass and its dynamic relationship with the surrounding stability.

This perspective reimagines gravity as a universal harmonizer. It binds objects into structured systems, from planetary orbits to galactic formations, while remaining fundamentally tied to the stability of the energy medium. By viewing gravity as an outcome of energy interactions rather than a purely mass-based attraction, we gain a deeper understanding of its true nature, unveiling its role as a balancing force in the cosmic tapestry.

8.2 Gravity: The Weakest Force in the Universe

Despite its ubiquitous presence and profound influence on the cosmos, gravity is fundamentally the weakest of the known forces in the universe. Compared to the electromagnetic force, strong nuclear force, and weak nuclear force, gravity's effects are almost negligible at the quantum or atomic scale. Yet, its significance lies not in its strength but in its reach and role in shaping the large-scale structure of the universe.

In the Unified Energy Dynamics (UED) framework, gravity's apparent weakness stems from its origin as a mass-energy interaction within the stable energy medium. Unlike the electromagnetic force, which operates through charged particles, or the strong nuclear force, which binds atomic nuclei, gravity arises from localized disturbances in the energy medium caused by mass. These disturbances ripple across the stable energy pool, creating the gravitational interactions that bind matter together.

Why Gravity Is Weak

Energy Distribution Across Scales: Gravity is diluted across vast cosmic distances, its effects spreading uniformly through the infinite energy medium. This contrasts with the localized and concentrated interactions of electromagnetic and nuclear forces.

Mass-Dependent Nature: Gravity depends solely on mass, which is often less concentrated in localized regions compared to the charge or nuclear properties that amplify other forces.

Stable Energy's Moderation: The stable energy pool acts as a universal regulator, absorbing and stabilizing gravitational disturbances. This moderation prevents gravity from overwhelming the delicate balance of the cosmos.

The Unique Role of Gravity

While gravity's weakness limits its influence at microscopic scales, it is the architect of large-scale cosmic structures. Its reach is unmatched, enabling it to shape galaxies, govern planetary orbits, and maintain the cohesion of galaxy clusters. Gravity's capacity to influence across vast distances ensures the universe remains interconnected and structurally coherent.

In the UED perspective, gravity is less of a force and more of a relational phenomenon—a silent but essential thread in the cosmic tapestry. Its weakness, far from being a limitation, allows it to operate harmoniously with other forces, enabling the universe to balance dynamism and stability across all scales.

The Energy Medium's Role in Gravitational Interactions

In the Unified Energy Dynamics (UED) framework, gravity is not an isolated force but a manifestation of interactions between mass and the stable energy medium. This reinterpretation shifts gravity from being merely a mass-based attraction to a phenomenon deeply connected with the energy medium's equilibrium. The stable energy pool, an infinite reservoir of potential energy, acts as the mediator for gravitational interactions, ensuring the universe's structural coherence and balance.

Stable Energy Pool: The Mediator of Gravity

Mass-Induced Disturbances: Gravity arises from disturbances created by mass within the stable energy medium. When mass interacts with this pool, it creates localized ripples or gradients that manifest as gravitational effects.

Energy Flow and Stability: The energy medium absorbs and stabilizes these disturbances, regulating the gravitational pull exerted by massive objects. This stabilization ensures that gravity remains consistent across scales, from planetary systems to galaxy clusters.

Equilibrium Maintenance: The stable energy pool functions as a balancing agent, preventing gravitational forces from becoming chaotic. This equilibrium is vital for maintaining the structural integrity of the universe.

The Relationship Between Gravity and Energy Medium Equilibrium

Dynamic Harmony: Gravity is an emergent phenomenon resulting from the interplay between active energy transitions and the stable reservoir. This dynamic ensures that mass-induced gravitational effects are not disruptive but harmonized within the energy medium.

Universal Connectivity: The stable energy pool connects all gravitational interactions, creating a cohesive network across the cosmos. This interconnectedness enables the gravitational pull of distant galaxies to influence one another, contributing to the large-scale structure of the universe.

Energy Medium as a Unifying Framework: Unlike other forces that operate through specific carriers like photons or gluons, gravity's influence is mediated directly by the stable energy pool. This makes it uniquely suited to operate across infinite distances, binding the universe into a unified whole.

Gravity's Unique Dependence on Stability

The stable energy medium not only mediates gravitational forces but also moderates them. Without this equilibrium, gravitational interactions could destabilize the cosmos, disrupting the intricate balance of forces that sustain galaxies, stars, and planetary systems. In this sense, gravity is not a standalone force but a harmonious interaction deeply rooted in the universal energy medium.

By understanding gravity as a product of mass-energy interactions within the stable pool, UED reveals its dual role: a subtle force shaping the cosmos and a vital element in maintaining the universe's equilibrium. This perspective underscores the fundamental connection between gravity and the infinite energy medium that sustains all existence.

8.3 Gravity Across Scales: From Particles to Galaxies

Gravity, as redefined within the Unified Energy Dynamics (UED) framework, is not constrained to a singular scale of existence. Its influence extends seamlessly from the subatomic realm of particles to the expansive structures of galaxies and beyond. This adaptability across scales reveals gravity's unique capacity to bridge the micro and macro realms, maintaining coherence in the universal energy medium.

Gravity in the Micro Realm

Quantum-Level Influence: While gravity is often overshadowed by the electromagnetic, strong, and weak nuclear forces at the quantum scale, it plays a subtle but essential role. Even in the interactions of particles like quarks and electrons, gravity's presence ensures a baseline stability within the universal energy medium.

Interplay with Energy Medium: At this scale, gravity manifests as a delicate distortion in the stable energy pool caused by the minuscule masses of particles. These distortions, though minimal, contribute to the foundational coherence of the quantum realm.

Unifying Active and Stable Energy: Gravity at the micro level underscores the interconnectedness of active energy transitions (like particle motion) and the stable energy reservoir, illustrating how even the smallest disturbances are harmonized within the energy medium.

Gravity in the Macro Realm

Galactic Structures and Beyond: On the cosmic scale, gravity's effects become dominant. It dictates the formation and motion of galaxies, stars, and planetary systems by channeling mass-induced disturbances through the stable energy medium.

Cosmic Filaments and Clusters: Gravity's large-scale influence creates the vast cosmic web of filaments and clusters, binding galaxies together while balancing the dispersive forces of expansion.

Shaping Celestial Bodies: The interaction of gravity with the stable energy medium governs phenomena such as orbital dynamics, planetary accretion, and the lifecycle of stars. It acts as the sculptor of celestial forms, ensuring their motion and evolution remain in harmony with universal principles.

Gravitational Waves and the Universal Energy Medium

Ripples in the Energy Pool: Gravitational waves, predicted by Einstein's General Relativity, are ripples in spacetime caused by massive, accelerating objects like merging black holes or neutron stars. UED reinterprets these waves as localized disturbances propagating through the stable energy medium.

Energy Dissipation and Stabilization: As gravitational waves traverse the cosmos, the stable energy medium absorbs and disperses their energy, maintaining equilibrium and preventing chaos. This interplay highlights the energy medium's role as both a transmitter and a stabilizer of gravitational effects.

Universal Communication: Gravitational waves serve as messengers, transmitting information about cosmic events across vast distances. Their interaction with the universal energy medium ensures their persistence and coherence, even over billions of light-years.

Gravity's Role in Cosmic Coherence

Gravity's ability to operate across scales—shaping the behavior of particles and the architecture of galaxies—makes it a unifying force within the cosmos. Its subtle influence at the micro level and dominant presence at the macro level exemplify the dynamic interplay between stability and motion, hallmarks of the universal energy medium.

By understanding gravity as an interaction mediated by the stable energy pool, UED reveals its true nature: a force that not only binds but also harmonizes. This perspective transforms gravity from a standalone phenomenon into a vital participant in the eternal dance of energy, bridging the smallest particles with the vastness of the universe.

8.4 Cosmic Harmony Through Gravitational Balance

Gravity, as the cornerstone of celestial mechanics, plays an indispensable role in maintaining the universe's equilibrium. Despite being the weakest of the fundamental forces, its long-range influence and universal presence make gravity a unifying force that sustains cosmic harmony. Within the framework of Unified Energy Dynamics (UED), gravity emerges not as an isolated phenomenon but as an integral player in the interplay of forces, energy, and matter.

Ensuring Equilibrium in Celestial Mechanics

Orbital Stability: Gravity governs the intricate dance of celestial bodies, ensuring their orbits remain stable over vast timescales. The gravitational pull between stars and planets creates balanced systems where motion follows predictable paths, such as the Earth's orbit around the Sun or the Moon's influence on Earth's tides.

Dynamic Balance of Forces: Celestial mechanics depends on the equilibrium between gravitational forces and other dynamics, such as inertial motion. This balance prevents planets from spiraling into stars or drifting aimlessly into space, enabling the formation of structured systems like galaxies and solar systems.

Binding Galactic Structures: At larger scales, gravity acts as the glue holding galaxies together. It counteracts the expansive force of the universe's ongoing expansion, allowing stars within galaxies to remain bound and contributing to the cohesion of galactic clusters.

The Synergy Between Gravity and Other Forces

Interplay with Electromagnetic Forces: Gravity complements electromagnetic forces in shaping the cosmos. While gravity governs the motion of massive bodies, electromagnetic interactions dominate at smaller scales, such as the behavior of charged particles in plasma clouds that give rise to stars.

Collaboration with Nuclear Forces: In the cores of stars, nuclear forces drive fusion reactions that generate immense amounts of energy. Gravity compresses these stellar cores, maintaining the conditions necessary for fusion and balancing the outward pressure of radiation with the inward pull of mass.

Mediating Expansion: Gravity's counteraction to the universe's expansion ensures that galaxies and cosmic structures do not dissolve into isolated components. This delicate interplay with the expansive tendencies of dark energy sustains the universe's large-scale organization.

Gravity as a Harmonizing Force

Dynamic Equilibrium: Gravity's influence ensures a state of dynamic equilibrium across scales, from the micro-level formation of planetary systems to the macro-level orchestration of galaxies. Its stabilizing effect harmonizes the chaos of energy transitions, fostering cosmic order.

Catalyst for Evolution: Gravity's ability to draw matter together acts as a catalyst for creation and transformation. It enables the coalescence of particles into planets, the birth of stars, and the formation of black holes, driving the universe's perpetual cycle of evolution.

Universal Connection: By acting as a mediator between mass and the energy medium, gravity connects disparate phenomena into a cohesive whole. It serves as a bridge between stability and motion, embodying the principles of balance that underpin the cosmos.

The Symphony of the Cosmos

Gravity, in its subtle yet pervasive presence, conducts the symphony of the universe. Through its intricate interactions with other forces and its integration with the stable energy medium, it ensures that the cosmos remains a harmonious and dynamic entity. Within the UED framework, gravity is more than a force—it is a universal balancer, preserving order amidst the ceaseless dance of energy and matter. This perspective invites us to view gravity not just as a mechanical phenomenon but as a vital thread in the tapestry of existence.

8.5 Revisiting Black Holes: Gravity's Extremes

Black holes, often regarded as enigmatic endpoints of stellar evolution, represent the extremes of gravitational interaction. Within the framework of Unified Energy Dynamics (UED), black holes are not merely regions of infinite density and inescapable gravity but key players in the universal cycle of energy transformation and dissolution. They highlight gravity's potential to shape and dissolve matter, offering profound insights into the nature of the energy medium and the role of singularities.

The Role of Gravity in Black Hole Formation

Collapse of Mass into Energy: Black holes form when the gravitational pull of a massive object overwhelms all opposing forces, causing it to collapse into an incredibly dense state. This process converts the object's mass into gravitational energy, pulling in nearby matter and energy.

Dissolution into Stability: Within the UED framework, black holes are seen as points where active energy and matter dissolve back into the stable energy pool. As matter crosses the event horizon, it undergoes a fundamental transformation, releasing active energy in the form of radiation or other phenomena, eventually reintegrating with the universal medium.

Cosmic Recycling: Black holes act as mechanisms for energy recycling, ensuring that the universe's dynamic equilibrium is maintained. They dissolve matter not into oblivion but into the infinite energy pool, continuing the cycle of creation and dissolution.

Insights into Gravitational Singularities

Localized Singularities, Not Universal Ends: Contrary to the classical interpretation of singularities as points of infinite density, UED views them as localized phenomena that reflect extreme disturbances in the stable energy medium. These singularities are not absolute but are bound by the principles of energy dynamics.

Gravitational Singularity and Energy Medium: At the heart of a black hole lies a singularity—a region where gravity's interaction with the energy medium becomes extreme. This interaction disturbs the equilibrium, leading to a concentrated release of active energy as matter transitions back to the stable reservoir.

Hawking Radiation as Transitional Energy: UED aligns with the concept of Hawking radiation, interpreting it as the emission of active energy during matter's dissolution. This radiation represents the ongoing interplay between gravity and the energy medium, bridging the gap between localized phenomena and the universal cycle.

Gravity and the Energy Medium at Extremes

Beyond Conventional Theories: While traditional physics struggles to reconcile quantum mechanics and relativity at black hole boundaries, UED provides a cohesive view. Gravity, as an interaction with the energy medium, operates seamlessly at these extremes, ensuring that even black holes adhere to the principles of energy dynamics.

A Gateway, Not an Endpoint: Black holes, within this perspective, are gateways where matter and energy transition rather than terminate. They exemplify the dynamic relationship between stability and motion, showcasing the versatility of the energy medium.

Black Holes as Catalysts of Understanding

Black holes, as the extremes of gravitational behavior, serve as laboratories for understanding the universe's most fundamental principles. They highlight gravity's profound role in shaping matter and energy, providing unique insights into the cyclical processes of creation and dissolution. Through the lens of UED, black holes emerge not as mysteries to be feared but as phenomena that embody the elegance of universal energy dynamics, bridging the finite and infinite realms of existence.

8.6 Conclusion: Gravity's Place in the Cosmic Tapestry

Gravity, often perceived as the weakest of forces, emerges in the framework of Unified Energy Dynamics (UED) as a profound binding agent that shapes the cosmos. Its role extends far beyond the classical understanding of an attractive force between masses, encompassing a dynamic interaction with the universal energy medium. Gravity bridges the tangible and intangible, orchestrating the balance between stability and motion that defines the universe.

Gravity as a Binding Force

Cosmic Sculptor: Gravity acts as the architect of celestial structures, shaping stars, galaxies, and clusters into harmonious patterns. It maintains the cohesion of matter within the vast expanse of space, preventing chaos while enabling cosmic order.

Dynamic Equilibrium: By interacting with the stable energy medium, gravity ensures the perpetual balance of energy and matter. Its influence stretches across scales, binding particles at the micro level and galaxies at the macro level into a unified cosmic tapestry.

Contributing to Universal Interplay

A Participant in Energy Dynamics: Gravity is not an isolated phenomenon but a participant in the universal energy cycle. It influences the flow of active energy, drives the formation and dissolution of matter, and contributes to the maintenance of equilibrium in the energy medium.

The Glue of Cosmic Harmony: Gravity's subtle strength binds the components of the universe together, enabling the dance of creation and dissolution. Its interplay with other forces and the energy medium ensures that the cosmos remains both dynamic and coherent.

A Unifying Perspective

Gravity's place in the cosmic tapestry is both humble and grand. While it may be the weakest force on a universal scale, its influence is foundational to the structure and function of the universe. UED redefines gravity as a bridge—connecting matter, energy, and the infinite medium of stability. In this light, gravity becomes a symbol of unity, a force that reminds us of the interconnected nature of all existence, and a testament to the elegance of the universe's design.

Summary of Part II

Time and Space in the Context of UED

1. Time: Human Perception in the Energy Medium

Traditional Approach: Time is perceived as a linear, universal constant, inherently tied to entropy and the irreversible progression of physical processes. Einstein's Relativity introduced time as a dimension intertwined with space, subject to the influences of gravity and velocity.

UED Approach: Time is a localized phenomenon, emerging from energy transitions and changes within physical systems. In the infinite stable energy pool, time holds no relevance, as the pool transcends temporal constraints and operates beyond linear progression. Human perception, shaped by entropy and the dispersal of active energy, creates the illusion of time's linearity, reflecting a subjective rather than universal truth.

2. The Nature of Space

Traditional Approach: Space is often regarded as an empty framework or vacuum, providing a stage for the placement and motion of objects. Modern physics incorporates the curvature of space as described by Einstein's General Relativity, wherein massive objects distort spacetime.

UED Approach: Space is the infinite, stable stage sustaining the interplay of active and stable energy, acting as the backbone of cosmic existence. Mass is not an inherent property but emerges as localized disturbances within this energy medium, shaping the dynamic structure of the cosmos. Space bridges the micro and macro realms, integrating all phenomena into a cohesive energy medium that underpins universal harmony.

3. The Role of Light and Perception

Traditional Approach: Light is treated as a constant in a vacuum, forming the basis for measuring cosmic distances (e.g., light-years). Photons are conceptualized as energy packets or wave-particle dualities, central to the theory of electromagnetism.

UED Perspective: Light is an emergent property of energy transitions, deeply rooted in the universal energy pool. While the speed of light appears constant in localized systems, it is not an absolute measure at the cosmic scale, given the dynamic and infinite nature of the energy medium.

Photons are transient manifestations of energy, emerging during transformations and dissolving back into the energy pool, symbolizing the dual nature of universal energy.

4. Gravity: Interactions and Cosmic Harmony

Traditional Approach: Gravity is viewed as a force of attraction between masses, explained through Newtonian mechanics or as spacetime curvature in Einstein's Relativity.

UED Perspective: Gravity emerges as an interaction between mass and the stable energy medium, responding to disturbances in the equilibrium of energy. Despite being the weakest force in terms of energy, gravity plays a critical role in ensuring cosmic harmony, balancing celestial mechanics across all scales. Gravitational extremes, such as black holes, serve as points where active energy reintegrates into the stable pool, completing the energy cycle.

5. Implications for Existing Frameworks

Revisiting Relativity and Quantum Mechanics: UED redefines foundational concepts like time, space, and gravity, offering a unified energy-based framework that bridges gaps between traditional models. The limitations of light-speed as a cosmic constant and gravitational calculations dependent on localized time zones are reconsidered within this new paradigm.

Reframing Cosmic Exploration: A need emerges for a speed-independent navigation system, leveraging energy dynamics rather than relying solely on relativistic constraints. UED introduces a perspective that aligns cosmic exploration with the principles of the infinite energy pool, potentially transforming interstellar navigation.

Conclusion of Part II

Part II extends the principles of Unified Energy Dynamics (UED) to redefine time, space, and gravity, challenging traditional frameworks and offering a cohesive vision where these constructs emerge from energy interactions within the universal medium. It repositions time as a localized perception, space as a stable yet dynamic stage, and gravity as a product of energy equilibrium rather than a standalone force. By aligning our understanding of the observable universe with the infinite potential of the energy pool, Part II opens pathways for innovative scientific exploration and a more profound grasp of cosmic harmony.

"What If?"

1. WHAT IF TIME DOESN'T Exist Beyond Human Perception?

Traditional Perspective: Time is a measurable dimension, essential for understanding the universe's structure, progression, and causality.

UED Perspective: If time is a construct tied to energy transitions and entropy, the universe may exist in a timeless state. This reimagines the notions of beginnings, endings, and the flow of events. Could timelessness be the true essence of the cosmos?

2. What If Light's Speed Isn't Universal?

Traditional Perspective: The speed of light is constant and foundational for calculating cosmic distances and spacetime behavior.

UED Perspective: If light's speed is influenced by localized energy dynamics and time zones, our understanding of cosmic distances could change dramatically. Could the universe be closer—or more expansive—than current measurements suggest?

3. What If Space Is Not a Vacuum but an Energy Medium?

Traditional Perspective: Space is conceptualized as an empty void, punctuated by matter and energy fields.

UED Perspective: If space is an infinite energy medium, it redefines celestial interactions, galaxy formation, and unexplained phenomena like dark matter or dark energy. Could this medium hold the key to understanding the universe's hidden architecture?

4. What If Gravity Isn't a Force?

Traditional Perspective: Gravity is an attractive force governed by mass and distance, explained by Newtonian mechanics or spacetime curvature.

UED Perspective: If gravity is an interaction between mass and the stable energy medium, it could explain galactic motion anomalies and black hole phenomena without invoking dark matter or exotic forces. Does this reframe gravity as a stabilizing interaction rather than a standalone force?

5. What If Distances in Space Are Miscalculated?

Traditional Perspective: Distances are calculated using light speed and localized time, assuming uniformity across the universe.

UED Perspective: If time and space are localized phenomena tied to energy dynamics, the concept of light-years could be fundamentally flawed. Could our maps of the universe require a complete re-evaluation?

6. What If Wormholes Aren't Shortcuts but Energy Voids?

Traditional Perspective: Wormholes are theoretical shortcuts through spacetime, requiring exotic matter for stability.

UED Perspective: If wormholes are voids in the energy medium rather than pathways, they could signify regions of universal equilibrium rather than instability. Could they serve as natural regulators of cosmic balance?

7. What If Black Holes Are Gateways to Stability?

Traditional Perspective: Black holes are regions of extreme gravity, absorbing matter and energy with no escape.

UED Perspective: If black holes reintegrate active energy into the stable pool, they could play an essential role in universal balance. Could they be viewed as cosmic recyclers rather than destructive endpoints?

8. What If the Universe Doesn't Expand but Redistributes Energy?

Traditional Perspective: The universe's expansion is evidenced by the redshift of galaxies and tied to the Big Bang theory.

UED Perspective: If the observed expansion is a redistribution of energy within the infinite medium, this could shift our understanding of the cosmos's origin and destiny. Could the energy pool's dynamics replace the need for an expansion-driven model?

9. What If Perception of Time Shapes Scientific Inquiry?

Traditional Perspective: Time is an objective parameter central to experiments and cosmological theories.

UED Perspective: If time is a human construct tied to observation, traditional time-dependent models may need revision. Could removing time as a central factor unlock new methods of scientific inquiry and discovery?

10. What If Energy, Not Space-Time, Is the True Unifier?

Traditional Perspective: Space-time is the universe's fabric, providing the foundation for relativity and unifying phenomena.

UED Perspective: If energy is the ultimate medium governing all phenomena, it could replace space-time as the universal framework. Does this shift revolutionize the search for a Theory of Everything and redefine the cosmos?

Conclusion

These refined "What If" scenarios challenge conventional thinking, encouraging readers to bridge traditional assumptions with UED's innovative insights. By questioning established frameworks and exploring UED's implications, these points invite deeper exploration of the universe's fundamental nature, inspiring curiosity and transformative inquiry.

Part III: Micro and Macro Energy Dynamics

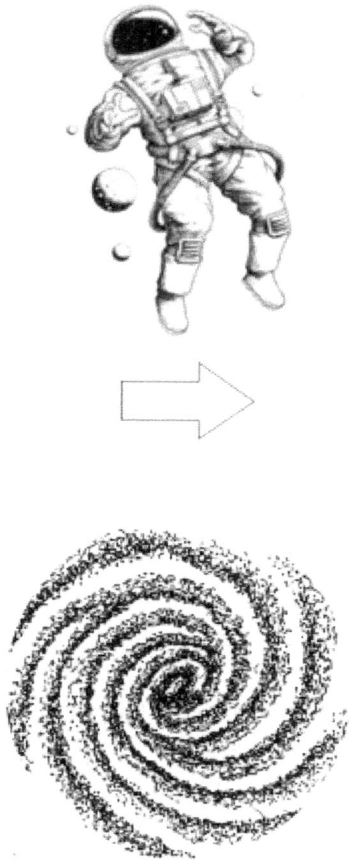

Chapter 9: The Photon and Energy Medium

Light has always held a mystical allure, weaving through human understanding as both a fundamental force of nature and a profound metaphor for knowledge and clarity. From ancient philosophies that revered it as the essence of divine creation to modern physics that unveiled its dual nature, light has remained central to our quest for understanding the universe. But what truly defines this enigmatic phenomenon? Within the framework of Unified Energy Dynamics (UED), the photon emerges not merely as a carrier of light but as the tangible bridge between stable and active energy—the universal messenger that connects the infinite energy pool with the observable cosmos.

This chapter embarks on a journey to decode the photon's role in the grand interplay of energy. We explore its unique properties, from wave-particle duality to its vibratory motion that sustains its existence as active energy. Through the lens of UED, photons become more than just packets of light; they are dynamic agents of transformation, revealing the unifying principles of the universe. As we unravel the photon's connection to the energy medium, we uncover the profound mechanisms through which it sustains the cosmic dance of creation and dissolution.

9.1 Introduction: Light as the Carrier of Energy

Light, the most familiar and enigmatic manifestation of energy, has fascinated humanity for centuries. At the heart of this phenomenon lies the photon, a particle that defies conventional categorization, embodying the dual nature of light as both wave and particle. In the framework of Unified Energy Dynamics (UED), photons emerge as more than just packets of light—they are the universal messengers of energy, bridging the stable energy pool with the active transformations that shape the cosmos.

Photons possess unique properties that make them integral to the understanding of energy dynamics. Their wave-particle duality, frequency-dependent energy, and interaction with the energy medium reveal profound insights into the nature of existence. They travel vast cosmic distances without a material medium, yet they remain intrinsically tied to the energy pool from which they arise. This interplay underscores their pivotal role in sustaining the universe's dynamic equilibrium.

In this chapter, we delve into the photon's essence, examining its properties through the lens of UED. By exploring its origin, behavior, and interactions, we aim to uncover how photons serve as carriers of energy, transforming our understanding of light and its place in the infinite energy medium.

9.2 The Dual Nature of Photons: Wave and Particle

The photon, a cornerstone of modern physics, is celebrated for its wave-particle duality—a property that has puzzled scientists and philosophers alike. Within the framework of Unified Energy Dynamics (UED), this duality is not merely a paradox but a reflection of the photon's unique role in the interplay between stable and active energy states.

Wave-Particle Duality Through Energy Dynamics

Wave-particle duality, traditionally explored in quantum mechanics, reveals that photons exhibit characteristics of both waves and particles, depending on the context of observation. In UED, this duality is seen as an emergent property of energy dynamics. The wave aspect of the photon represents its active connection to the stable energy medium, where it propagates as oscillating electromagnetic fields. The particle aspect, on the other hand, signifies the localized nature of energy when interacting with matter, such as in the photoelectric effect or atomic transitions.

This duality highlights the adaptability of photons, allowing them to act as bridges between the stable reservoir of energy and the dynamic transformations of the physical universe. As waves, they maintain the vibrational continuity of energy. As particles, they deliver discrete quanta of energy to systems, enabling specific transitions and transformations.

Photons as Energy Transitions

Photons embody the process of energy transitioning between stable and active states. They emerge from disturbances in the stable energy pool, such as those caused by atomic transitions, nuclear reactions, or thermal emissions. During their journey, they carry energy in the form of electromagnetic waves, interacting with various systems to drive processes like heating, excitation, or chemical reactions.

Their dual nature ensures that energy transitions are seamless, preserving the connection between the infinite energy medium and the finite manifestations of active energy. By oscillating between wave-like propagation and particle-like interaction, photons demonstrate the fluidity and interconnectedness of energy dynamics, serving as tangible evidence of the universe's vibrational harmony.

Through this lens, photons are not merely particles of light but carriers of universal energy, integral to the dance of creation and transformation that defines existence itself.

9.3 Vibratory Motion: The Signature of Active Energy

Vibratory motion is the heartbeat of active energy, manifesting as the rhythmic oscillations that sustain its dynamic nature. For photons, these vibrations are not just a byproduct of motion but the very mechanism through which they maintain their form and interact with the universal energy medium. Within Unified Energy Dynamics (UED), vibratory motion emerges as a defining characteristic of active energy, connecting it to the stable energy pool and facilitating its role as a carrier of transformation.

Sustaining Photons Through Vibratory Motion

Photons, as embodiments of energy in motion, rely on vibratory motion to sustain their identity as packets of electromagnetic energy. This motion is expressed through the oscillation of electric and magnetic fields, which are fundamental and universal components of energy dynamics. These fields oscillate perpendicular to each other and to the direction of the photon's propagation, creating a self-sustaining wave that requires no material medium to traverse space.

This vibratory nature ensures that photons retain their active energy state while maintaining their connection to the stable energy pool. The oscillating fields are, in essence, the photon's "wings," enabling it to move seamlessly through the energy medium, carrying energy across vast distances without losing its identity or purpose.

Frequency: The Universal Property of Active Energy

Frequency, the measure of how often a photon's electromagnetic fields oscillate in a given time frame, serves as a universal signature of its energy. According to UED, frequency is more than a physical attribute—it is the bridge linking the photon's active energy to the infinite stable energy pool.

The constancy of a photon's frequency, regardless of the medium it traverses, underscores its connection to the universal energy medium. Even as photons interact with matter or traverse diverse environments, their frequency remains unaltered, signifying the invariance of energy's fundamental nature. This constancy also allows frequency to act as a precise indicator of a photon's energy level, aligning with the well-established relationship $E=h\nu$, where E is energy, h is Planck's constant, and ν\ nuν is frequency.

The Vibratory Connection to the Energy Medium

The vibratory motion of photons is not an isolated phenomenon but a reflection of the inherent vibrational nature of energy itself. Whether in matter, where atomic and molecular vibrations sustain active energy, or in photons, where field oscillations propagate energy, vibratory motion serves as the universal mechanism that ensures coherence and continuity within the energy medium.

By maintaining their vibratory motion, photons embody the active state of energy, constantly interacting with and transferring energy within the cosmos. This motion reinforces their role as conveyors of universal energy, harmonizing the micro and macro realms into a unified dynamic.

In the grand interplay of the universe, vibratory motion emerges as the signature of active energy, encapsulating its dynamic essence while maintaining its eternal connection to the stable energy pool.

It is this rhythmic dance that enables photons to sustain their journey, delivering energy and bridging the tangible and intangible aspects of existence.

9.4 Electromagnetic Fields: The Wings of Photons

Electromagnetic fields are the unseen wings that enable photons to soar through the universe, carrying energy across vast expanses of space. These oscillating electric and magnetic fields are not merely components of light but fundamental elements of energy dynamics, allowing photons to sustain their motion and transfer energy without the need for a material medium. Within Unified Energy Dynamics (UED), electromagnetic fields represent a universal principle that bridges the tangible and intangible realms of existence.

Electric and Magnetic Fields: Enablers of Photon Propagation

The propagation of photons is driven by the interplay of electric and magnetic fields, which oscillate perpendicularly to each other and to the direction of the photon's motion. This self-sustaining configuration allows photons to move through space without requiring a physical medium, unlike mechanical waves that depend on particle vibrations within a material substance.

Electric Field Oscillations: The electric field of a photon alternates in magnitude and direction, creating a rhythmic wave that influences charged particles and interacts with matter.

Magnetic Field Oscillations: Perpendicular to the electric field, the magnetic field oscillates in harmony, completing the electromagnetic wave. This dual-field structure ensures the stability and propagation of photons as they traverse the energy medium.

These oscillating fields are intrinsic to the photon's nature, enabling it to transport energy effectively and interact with other forms of matter and energy.

Universality of Electromagnetic Fields in Energy Dynamics

Electromagnetic fields are not confined to photons alone; they are a universal feature of energy interactions across all scales. From the radiation emitted by stars to the magnetic fields surrounding planets, these fields play a pivotal role in shaping the cosmos:

Cosmic Scale: Electromagnetic waves generated by celestial events, such as supernovae and quasars, propagate through the energy medium, carrying information and energy across intergalactic distances.

Quantum Scale: At the subatomic level, electromagnetic interactions govern the behavior of charged particles, such as electrons and protons, influencing chemical bonding and atomic stability.

Within the UED framework, electromagnetic fields are seen as extensions of the stable energy medium, facilitating energy transitions and sustaining the vibratory motion of active energy.

Electromagnetic Fields and the Energy Medium

The interaction of electromagnetic fields with the stable energy medium highlights their universality and significance. Unlike mechanical waves that rely on particle vibrations, photons interact directly with the energy medium, using their oscillating fields to maintain motion and coherence. This interaction underscores the integral role of electromagnetic fields as carriers of energy transitions, connecting the micro and macro realms in a unified dynamic.

In the context of UED, electromagnetic fields emerge as the true "wings" of photons, allowing them to navigate the universe as messengers of energy. These fields ensure that photons remain connected to the stable energy pool, embodying the active state of energy while sustaining the flow of creation, transformation, and dissolution.

Electromagnetic fields are more than a mechanism for photon propagation—they are a universal principle that underpins energy dynamics. Through their oscillations, they reveal the interconnectedness of the cosmos, harmonizing the tangible and intangible aspects of existence into a coherent and infinite whole.

9.5 Photon Emission and Dissolution: The Cycle of Energy

Photons, as universal messengers of energy, play a vital role in the perpetual cycle of creation, transformation, and dissolution within the cosmos. They emerge from dynamic energy transitions and eventually reintegrate into the stable energy pool, embodying the harmony and continuity of the universe. This process, central to Unified Energy Dynamics (UED), highlights the interconnected nature of all phenomena, governed by the timeless principles of energy.

Emergence of Photons from Energy Transitions in Matter

Photons are born from the interaction of active energy within matter, serving as carriers of energy released during various transitions. These emissions occur across a spectrum of processes, ranging from atomic and molecular scales to celestial events:

Atomic and Molecular Emissions: When electrons in an atom transition between energy levels, they emit photons corresponding to the energy difference between the states. This process, observed in phenomena like fluorescence and atomic spectra, showcases the precise and quantized nature of photon emission.

Thermal Radiation: As objects are heated, the active energy within them manifests as electromagnetic radiation, emitting photons across a spectrum of wavelengths. This process explains the glow of heated metals and the thermal emissions from stars.

Nuclear Reactions: In nuclear fission and fusion, immense amounts of active energy are released, generating photons in the form of gamma radiation. This energy release powers stars and fuels the universe's dynamic processes.

Celestial Events: Cosmic phenomena like supernovae and quasars release vast amounts of energy in the form of photons, illuminating the universe and propagating energy across intergalactic distances.

Through these transitions, photons emerge as visible and detectable manifestations of active energy, bridging the tangible and intangible realms of existence.

Reintegration of Photons into the Stable Energy Pool

The life cycle of photons does not end with their emission; their eventual dissolution represents a return to the universal energy pool, maintaining equilibrium within the cosmos. This reintegration occurs as photons interact with matter or encounter extreme conditions:

Absorption by Matter: When photons interact with matter, their energy is transferred to the absorbing material, fueling processes like photosynthesis, photoelectric effects, or heat generation. This absorption marks the dissolution of photons into localized energy expressions.

Energy Dissipation in Space: Photons traveling vast cosmic distances may lose energy through redshift or scattering, gradually dispersing their active state and merging back into the stable energy medium.

Extreme Conditions: In black holes, photons are drawn into regions of intense gravitational influence, where their energy transitions back into the stable pool, exemplifying the cycle of creation and dissolution at the universe's extremes.

The Universal Cycle of Energy

This cyclical process of photon emission and dissolution reflects the balance and harmony of energy dynamics within UED. Photons, as transient forms of active energy, embody the dynamic interplay between stability and motion:

- ✓ **Creation**: Energy transitions give rise to photons, allowing the transfer and transformation of energy across systems.

- ✓ **Propagation**: Photons traverse the cosmos, interacting with the energy medium and delivering their energy payload.

- ✓ **Reintegration**: Photons dissolve back into the stable energy pool, ensuring continuity and equilibrium within the infinite medium.

In this cycle, photons serve as vital connectors, linking the micro and macro realms, the tangible and the abstract, and the transient and the eternal. By understanding this cycle, we gain deeper insights into the universe's perpetual motion, where energy is neither created nor destroyed but endlessly transformed, sustaining the cosmic dance of existence.

9.6 Photon Dynamics across Media

Photons, as the carriers of electromagnetic energy, possess unique properties that enable their propagation through both material and non-material realms. Unlike mechanical waves, which rely on the vibration of particles in a medium, photons traverse the universe without the need for a material medium. Their behavior reflects their interaction with the energy medium itself, as well as their interactions with confined active energy in various systems.

Why Photons Propagate Without Requiring a Material Medium

The propagation of photons through space, independent of any material substrate, is one of the most intriguing phenomena of energy dynamics. This unique property is a result of their inherent nature as oscillating electromagnetic fields:

Energy Medium as the Universal Framework: Photons do not require a material medium because their propagation is rooted in the energy medium itself. This infinite, stable energy pool serves as the universal framework that supports the oscillation of electromagnetic fields. The energy medium provides the stability and coherence needed for photons to carry energy across vast distances.

Self-Sustained Oscillations: Photons consist of oscillating electric and magnetic fields that perpetuate each other. These fields are mutually perpendicular and self-sustaining, allowing the photon to propagate through the energy medium without needing physical particles to transmit their energy.

Independence from Material Constraints: Unlike mechanical waves, whose speed and behavior depend on the properties of the material medium, photons propagate at a constant speed in a vacuum, defined by the universal energy medium. This independence highlights the photon's unique connection to the stable energy pool.

Interaction of Photons with Confined Active Energy in Various Systems

While photons propagate freely through the energy medium, their interactions with confined active energy reveal their dynamic role in energy transitions. These interactions vary across different systems and conditions:

In Transparent Media: When photons pass through transparent materials like glass or water, their speed decreases due to interactions with the material's active energy. This delay occurs as photons temporarily excite electrons in the medium, transferring energy before continuing their journey. However, their frequency remains constant, ensuring the conservation of their energy signature.

In Absorptive Systems: In systems like photosynthetic cells or photoelectric materials, photons transfer their energy to confined active energy, initiating processes like electron excitation or chemical reactions. This interaction transforms photon energy into localized forms, showcasing their role in driving energy transitions.

In High-Energy Environments: In conditions like nuclear reactions or stellar cores, photons emerge from the active energy released during transformations. For example, gamma rays produced in nuclear fusion carry immense energy, reflecting the dynamic interplay between photons and confined active energy.

In Cosmic Conditions: Photons traversing the cosmos may interact with interstellar and intergalactic matter, scattering and redshifting as they exchange energy with confined active systems. These interactions provide valuable insights into the structure and behavior of the universe.

The Photon as a Bridge in Energy Dynamics

Through their propagation and interactions, photons act as bridges between stable energy and active phenomena. Their ability to navigate the infinite energy medium and interact with matter makes them essential components of the universe's energy framework:

Connecting Micro and Macro Scales: Photons carry information and energy across vast distances, linking quantum-scale processes with cosmic-scale phenomena.

Driving Transformations: By transferring energy to and from confined systems, photons enable the dynamic processes that sustain life, drive celestial events, and maintain cosmic balance.

Reflecting Universal Principles: The behavior of photons across media highlights the universality of energy dynamics, governed by the interplay between stability and motion.

In the context of Unified Energy Dynamics (UED), photons exemplify the harmonious interaction between the energy medium and active energy. Their propagation without a material medium and their transformative interactions with confined energy systems reaffirm their role as fundamental conveyors of energy, sustaining the perpetual dance of creation, transformation, and dissolution.

9.7 Frequency and Energy: The Constant Connection

The relationship between frequency and energy is one of the most remarkable and consistent aspects of photon behavior. Across space and through various media, the frequency of a photon remains invariant, acting as a fundamental property that defines its energy and interaction with the universe. This constancy reflects deeper principles within Unified Energy Dynamics (UED), revealing the significance of frequency as a universal identifier of active energy.

Why Frequency Remains Invariant Across Space and Media

Frequency, the number of oscillations a wave undergoes per second, is an intrinsic property of photons that does not change during their journey. Unlike wavelength, which can vary depending on the medium and conditions of propagation, frequency remains constant, ensuring that the photon retains its energy signature. This invariance is rooted in the following principles:

Energy Conservation: According to UED, the energy of a photon, defined by E=hνE = h \nuE=hν (where hhh is Planck's constant and ν\nuν is frequency), remains constant unless it undergoes a transformative interaction. Even as photons pass through different media and their speed and wavelength adjust, their frequency stays fixed to preserve the conservation of energy.

Independence from Medium Properties: While photons interact with confined active energy in different media, these interactions affect their propagation speed and wavelength rather than their frequency. This constancy ensures that photons maintain their identity and energy during transitions.

Universality of Vibratory Motion: Frequency reflects the inherent vibratory motion of active energy within the photon. This vibratory nature ties the photon directly to the stable energy medium, transcending localized disruptions or variations caused by external factors.

The Significance of Frequency in Determining Photon Energy and Universal Properties

Frequency is not just a measure of oscillations; it is the defining characteristic that connects photons to the universal energy framework. Its significance extends across multiple dimensions:

Defining Photon Energy: The energy of a photon is directly proportional to its frequency, making frequency the key determinant of its energy level. High-frequency photons, such as gamma rays, carry immense energy, while low-frequency photons, like radio waves, exhibit less energetic states. This proportionality underscores the importance of frequency as a universal measure of active energy.

Facilitating Universal Interactions: Photons interact with matter and the energy medium based on their frequency.

For example: High-frequency photons can excite atomic nuclei, triggering nuclear reactions or emission phenomena.

Medium-frequency photons, like visible light, interact with electrons in matter, producing phenomena such as the photoelectric effect or photosynthesis.

Low-frequency photons influence larger-scale systems, such as communication technologies and atmospheric dynamics.

Frequency as a Signature of Origin: The frequency of a photon often reveals its source, whether it emerges from atomic transitions, thermal radiation, or cosmic processes.

For instance: The cosmic microwave background radiation exhibits a characteristic frequency, serving as a remnant signature of early energy transitions in the universe.

Stellar spectra reveal distinct frequencies corresponding to the energy states of emitting atoms, providing insights into a star's composition and temperature.

Universal Connectivity: Frequency acts as a bridge between micro and macro realms, linking quantum-scale interactions with cosmic-scale phenomena. Through their frequency, photons embody the universality of energy transitions, reflecting the interconnectedness of the cosmos.

Frequency in the Context of UED

Unified Energy Dynamics positions frequency as the signature property of active energy, a constant that ties photons to the infinite energy medium:

Active Energy in Motion: Frequency reflects the vibratory nature of active energy, enabling photons to sustain their dynamic state while maintaining a connection to the stable energy pool.

Key to Energy Transitions: By maintaining a constant frequency, photons ensure the continuity and coherence of energy transformations across systems, from atomic to astronomical scales.

Indicator of Universal Principles: The invariance of frequency highlights the underlying order within the energy medium, reinforcing the stability and harmony of cosmic processes.

Frequency is more than a numerical value—it is the heartbeat of energy in motion, the constant that allows photons to navigate the infinite energy medium and sustain the dance of creation and transformation. Within UED, frequency emerges as the ultimate marker of active energy, bridging the realms of the observable and the universal with unparalleled consistency.

9.8 The Photon as Evidence of Universal Energy

Photons, with their unique properties and behaviors, stand as compelling evidence of the universal energy principles outlined in Unified Energy Dynamics (UED). From their wave-particle duality to their interactions across scales, photons embody the seamless interplay between active and stable energy. By examining their role in the cosmos, we uncover profound insights into the mechanisms that sustain universal harmony.

How Photon Behavior Confirms the Principles of UED

The characteristics and actions of photons align closely with the foundational concepts of UED, providing observable proof of the energy medium's dynamic and stable nature:

Wave-Particle Duality and Energy States: Photons exhibit dual behavior as waves and particles, a phenomenon that reflects their role as active energy in transit. This duality mirrors the interplay of stability and dynamism within the energy medium, with photons acting as vehicles for energy transformations.

Emergence from Energy Transitions: Photons arise from energy transitions within matter, such as atomic or nuclear processes, underscoring their origin in the interaction between stable and active energy.

For example: In stars, photons are released as nuclear reactions convert mass into radiant energy, demonstrating the cycle of energy emergence and transformation.

In quantum systems, photons appear during electron transitions, linking their behavior to the vibratory nature of active energy.

Constancy of Frequency and Energy Transfer: A photon's frequency remains invariant, reflecting the continuity and coherence of energy transitions. This invariance aligns with UED's principle that energy transformations maintain equilibrium within the universal medium.

Photon Dissolution and Reintegration: When photons encounter black holes or other extreme conditions, they dissolve back into the stable energy pool, completing the cycle of creation and dissolution. This reintegration illustrates UED's concept of the energy medium as both source and sink, maintaining cosmic balance.

Insights into the Photon's Role in Cosmic Harmony

Photons are not merely carriers of energy; they are essential contributors to the interconnected fabric of the universe. Their behavior illuminates their role in sustaining cosmic harmony:

Connecting Micro and Macro Realms: Photons bridge the quantum and cosmic scales, transferring energy across vast distances and enabling interactions between particles, atoms, and celestial bodies. This universality underscores their significance in unifying disparate aspects of existence.

Facilitating Communication Across the Cosmos: As electromagnetic waves, photons act as messengers of the universe, carrying information from one region of space to another. From the light of distant stars to the cosmic microwave background, photons serve as the universe's storytellers, revealing its history and dynamics.

Driving Energy Transformations: Photons play a central role in energy redistribution, whether through the radiation of heat, the emission of light, or the propagation of electromagnetic fields. This continuous movement ensures that the energy medium remains dynamic and balanced.

Symbols of Universal Order: The predictable behavior of photons, governed by constants like Planck's constant and the speed of light, reflects the underlying order of the cosmos. Their actions reaffirm UED's assertion that the universe operates as a cohesive and harmonious system.

The Photon in the Context of UED

Under the Unified Energy Dynamics framework, photons emerge as tangible evidence of the energy medium's universal principles:

Active Energy in Transit: Photons embody the vibratory motion of active energy, connecting the stable pool to observable phenomena.

Messengers of Equilibrium: By transferring energy across systems and facilitating transitions, photons maintain the balance between stability and motion, ensuring cosmic harmony.

Indicators of Universal Connectivity: The photon's role in unifying the micro and macro realms highlights the interconnected nature of existence, a cornerstone of UED.

In the vast expanse of the universe, photons serve as the dynamic threads that weave the fabric of reality. Their behavior, consistent with UED's principles, offers profound insights into the nature of energy and its role in shaping the cosmos. By understanding photons within this framework, we gain not only a deeper appreciation for their properties but also a clearer vision of the universal energy medium that sustains all existence.

9.9 Conclusion: Photons as Universal Messengers

Photons, as the carriers of light and energy, hold a profound position in the grand tapestry of the universe. They embody the fundamental principles of Unified Energy Dynamics (UED), serving as the bridge between stable and active energy states. This duality makes photons not just particles of light but universal messengers that convey the essence of energy's infinite dance.

Reaffirming Photons' Role in Bridging Stable and Active Energy

Photons emerge as vibrant expressions of the energy medium, illustrating the dynamic interplay between stability and motion. Their wave-particle duality reflects the seamless transition of energy from the stable pool into active forms, showcasing the infinite versatility of the energy medium. Whether in nuclear reactions, electromagnetic propagation, or cosmic emissions, photons exemplify how active energy manifests, interacts, and ultimately returns to its source.

Through their interactions, photons illuminate the intrinsic connection between energy's stable and dynamic states. Their constant frequency, invariant across the cosmos, underscores their role as stabilizers in a perpetually transforming universe. By maintaining coherence and continuity, photons uphold the balance that defines the universal energy framework.

How Photons Embody the Dynamic Equilibrium of the Cosmos

Photons are the tangible evidence of energy's cyclical nature, emerging from energy transitions and dissolving back into the stable reservoir. Their behavior—whether transferring energy across vast distances or enabling quantum-level interactions—reaffirms the universe's harmony. They are not just carriers of energy but custodians of cosmic equilibrium, ensuring that transformation and stability coexist seamlessly.

Photons' universality bridges scales, from the microscopic realm of quantum particles to the vastness of intergalactic space. As they traverse the cosmos, they carry with them the story of energy's eternal motion, uniting distant regions of space and enabling the interconnectedness that defines existence.

In the context of UED, photons are far more than mere particles of light; they are the dynamic threads that weave the universe into a cohesive whole. Their journey from emergence to reintegration reflects the timeless principles of energy dynamics, offering a glimpse into the intricate and harmonious nature of reality. By understanding photons as universal messengers, we embrace a perspective that transcends traditional boundaries, uncovering the profound interconnectedness of all that exists.

Chapter 10: The Mechanics of Waves in the Energy Medium

As we delve deeper into the dynamics of energy in its infinite medium, we arrive at a phenomenon that bridges the micro and macro realms: waves. Waves are the universal language of energy, facilitating its transfer and transformation across diverse systems. From the gentle ripples of water to the intricate oscillations of electromagnetic fields, waves embody energy in motion. They connect particles, fields, and cosmic structures, serving as the carriers of information and interaction throughout the universe.

Building upon our understanding of photons and their role as messengers of energy, this chapter explores the broader mechanics of wave propagation in the energy medium. How do electromagnetic waves traverse vast distances without a material medium? What distinguishes them from mechanical waves, which require vibrating particles to propagate? And how does the stable energy pool influence the behavior of waves, from the quantum scale to interstellar phenomena? By examining these questions, we uncover the profound harmony that waves bring to the cosmos, guided by the principles of Unified Energy Dynamics (UED).

10.1 Introduction: Waves as Energy in Motion

Waves are the quintessential expression of energy in motion, weaving through the universe as both carriers and manifestations of energy transitions. Whether rippling through a pond, traversing the vastness of space as light, or vibrating through matter as sound, waves embody the dynamic interplay of stability and motion within the energy medium. They are the threads that connect the tangible and intangible, revealing the mechanisms by which energy propagates and interacts across scales.

Unified Energy Dynamics (UED) offers a transformative lens to understand waves, not merely as phenomena within isolated systems but as fundamental processes tied to the infinite energy pool. This chapter examines how waves emerge, propagate, and influence the fabric of existence, setting the stage for a deeper exploration of their universal role. By revisiting the duality of electromagnetic waves, comparing them with mechanical waves, and analyzing their behavior within the stable energy medium, we uncover the principles that govern energy's journey through the cosmos.

10.2 Electromagnetic Waves in the Energy Medium

Electromagnetic waves are one of the most profound manifestations of energy in motion, characterized by oscillating electric and magnetic fields that propagate through space. Unlike mechanical waves, which require a material substrate for their propagation, electromagnetic waves traverse the cosmos independently of traditional matter-based media. This distinction makes them uniquely suited to explore the fabric of the universe, especially through the lens of Unified Energy Dynamics (UED).

The Nature of Electromagnetic Waves: Fields in Oscillation

Electromagnetic waves consist of two intertwined fields—electric and magnetic—that oscillate perpendicularly to each other and the direction of wave propagation. These waves arise from the dynamic interactions of charged particles, where changes in an electric field generate a magnetic field, and vice versa. This interplay creates a self-sustaining wave capable of traveling vast distances without attenuation in the absence of matter.

In UED, these oscillating fields are not merely physical phenomena but expressions of active energy within the stable energy medium. Each wave is a localized disturbance that represents energy transitioning through space, a testament to the dynamic equilibrium maintained within the universal energy pool.

The Role of the Energy Medium in Wave Propagation

While traditional physics posits that electromagnetic waves require no medium for propagation, UED reinterprets this phenomenon by introducing the concept of the infinite energy pool as the universal medium. This stable energy reservoir provides the framework within which these waves travel, ensuring coherence and continuity even in the vacuum of space.

The energy medium allows electromagnetic waves to propagate by supporting the oscillatory dynamics of their fields. Rather than relying on the vibration of particles, as in mechanical waves, electromagnetic waves interact directly with the stable energy pool, leveraging its inherent properties to maintain their motion. This interaction explains why electromagnetic waves can traverse the emptiest regions of space, from intergalactic voids to the vast expanses between stars.

Through this lens, the energy medium becomes an active participant in the propagation of electromagnetic waves, facilitating their journey across the cosmos and underscoring their role as conveyors of energy transitions. By understanding this relationship, we gain deeper insights into the fundamental mechanisms that govern the flow of energy within the universe.

10.3 The Contrast with Mechanical Waves

Mechanical waves and electromagnetic waves represent two distinct modes of energy transfer, each governed by unique principles and interactions with their respective mediums. By examining the contrast between these two types of waves, we can better appreciate the underlying mechanics of energy propagation and the transformative role of the energy medium in Unified Energy Dynamics (UED).

Mechanical Waves: Vibrations in a Material Medium

Mechanical waves, such as sound, water, and seismic waves, depend entirely on the presence of a material medium for their propagation. They transfer energy through the vibratory motion of particles within the medium, whether solid, liquid, or gas. For instance:

✓ **Sound waves** rely on compressions and rarefactions in air molecules to travel.

✓ **Water waves** propagate as the oscillatory motion of water particles transfers energy across the surface.

✓ **Seismic waves** move through the Earth's crust via the displacement of solid particles.

In each case, the energy of the wave is carried by the coordinated motion of particles in the medium, but these particles themselves do not migrate with the wave.

Electromagnetic Waves: Propagation Without a Material Medium

In stark contrast, electromagnetic waves require no physical medium for their propagation. Instead, they are self-sustaining oscillations of electric and magnetic fields, traveling through space at the speed of light. This independence from material substrates allows electromagnetic waves to traverse even the emptiest regions of the universe, such as the vacuum of space.

Through the UED framework, this phenomenon is reinterpreted: electromagnetic waves interact with the stable energy pool, the universal medium that transcends traditional matter. Rather than relying on particle vibrations, these waves draw from the inherent properties of the energy medium, enabling their seamless motion through space.

Comparing the Role of the Medium

The fundamental difference lies in how the medium facilitates energy transfer:

Mechanical Waves: The medium actively participates in the energy transfer process through the vibratory motion of its particles. Without a material medium, mechanical waves cannot propagate.

Electromagnetic Waves: The energy medium provides the framework for wave propagation but does not rely on particle interactions. Instead, it supports the oscillatory dynamics of electric and magnetic fields, maintaining wave motion across any environment, even a vacuum.

This comparison underscores the unique nature of electromagnetic waves as manifestations of energy transitions within the stable energy pool. While mechanical waves illustrate energy transfer through localized particle interactions, electromagnetic waves reveal the universal interplay between active energy and the energy medium, highlighting the profound dynamics that sustain energy flow across the cosmos.

10.4 The Role of Energy Medium in Wave Propagation

The propagation of waves, whether mechanical or electromagnetic, is a direct manifestation of energy transitions. In the framework of Unified Energy Dynamics (UED), the energy medium—comprising the stable energy pool—plays a pivotal role in facilitating this motion. By interacting with wave fields, the energy medium not only enables the transfer of energy but also ensures coherence and continuity across all scales, from quantum fluctuations to cosmic phenomena.

The Energy Medium as a Dynamic Support

Unlike traditional interpretations that treat space as a passive vacuum, UED posits that the stable energy medium is an active participant in wave propagation:

Electromagnetic Waves: The stable energy pool interacts with oscillating electric and magnetic fields, providing the framework that sustains their motion. This interaction ensures the self-propagating nature of electromagnetic waves, which can traverse the vastness of space without a material substrate.

Mechanical Waves: While mechanical waves depend on a material medium for particle vibrations, the energy medium underpins the stability of this material, maintaining the conditions necessary for mechanical wave propagation.

The energy medium serves as a bridge, connecting the dynamics of localized phenomena with the universal flow of energy.

Energy Transfer Mechanisms in the Infinite Medium

The process of wave propagation involves the transfer of energy without the physical displacement of the medium itself. In UED, this is facilitated by the stable energy pool through the following mechanisms:

Field Oscillations in Electromagnetic Waves: The stable energy medium supports the oscillations of electric and magnetic fields, allowing energy to move through space at the speed of light. This interaction ensures that photons, as packets of active energy, maintain their vibratory motion while transferring energy across the cosmos.

Particle Vibrations in Mechanical Waves: For mechanical waves, the energy medium stabilizes the material particles, enabling them to oscillate in response to applied forces. The energy medium ensures that these vibrations are coherent and capable of transmitting energy efficiently.

Universal Coherence: The energy medium maintains the harmony of wave interactions across scales. Whether it is the quantum-level vibrations of particles or the macro-scale propagation of gravitational waves, the energy medium ensures that the principles of energy transfer are universally consistent.

A Medium That Transcends Boundaries

The energy medium's interaction with wave fields reflects its fundamental role as the universal substrate of existence. Its stability accommodates the dynamic properties of waves, allowing energy to manifest as motion, vibration, and oscillation. By providing a consistent framework for energy transfer, the energy medium unifies the behavior of waves across all forms and scales, ensuring that the cosmos remains interconnected and harmonious.

Through this lens, wave propagation is not merely a localized phenomenon but a universal process driven by the interplay between active energy and the infinite stable energy pool. This perspective not only bridges the gap between mechanical and electromagnetic waves but also redefines our understanding of how energy flows through the fabric of existence.

10.5 Cosmic Wave Dynamics and the Energy Medium

The vast expanse of the cosmos is a dynamic arena where waves of various kinds propagate, transferring energy across unimaginable distances. In the framework of Unified Energy Dynamics (UED), these waves—whether electromagnetic, mechanical, or gravitational—are deeply influenced by the stable energy medium. The energy medium serves as the universal substrate, guiding and sustaining these waves while ensuring coherence in the vast and complex interactions of the cosmos.

The Energy Medium: Shaping Cosmic Waves

The energy medium plays a fundamental role in defining the behavior of cosmic waves. Its stability provides the foundation for energy propagation, while its interaction with active energy ensures that waves maintain their coherence and dynamics across scales.

Electromagnetic Waves in Cosmic Space: Electromagnetic waves, from radio signals to gamma rays, traverse vast interstellar and intergalactic distances. In the absence of a material medium, the stable energy pool acts as the backdrop that supports the oscillating electric and magnetic fields, allowing photons to propagate seamlessly. This interaction is what enables electromagnetic waves to carry energy and information across the universe, unaffected by the lack of a traditional medium.

Mechanical Wave Constraints: Unlike electromagnetic waves, mechanical waves are restricted to material media. The energy medium stabilizes the particles within these materials, enabling vibrations and energy transfer. While mechanical waves have limited reach in cosmic settings due to the sparsity of matter, their behavior within dense regions, like planetary interiors or stellar surfaces, is governed by the energy medium's stabilizing influence.

Gravitational Waves: Ripples in the Universal Pool

One of the most fascinating manifestations of wave dynamics in the cosmos is gravitational waves. Predicted by Einstein's General Relativity and later observed experimentally, gravitational waves are distortions in spacetime caused by massive cosmic events, such as merging black holes or colliding neutron stars. From the perspective of UED, these waves can be reinterpreted as ripples in the stable energy pool.

Origin of Gravitational Waves: When massive objects disturb the energy equilibrium of the cosmos—whether through collisions, rapid accelerations, or other dynamic events—they create fluctuations in the stable energy pool. These disturbances propagate outward as gravitational waves, carrying information about the event's scale, intensity, and nature.

Propagation Through the Energy Medium: Gravitational waves interact with the energy medium much like ripples on a pond. The stable energy pool provides the coherence that allows these waves to travel vast distances without significant dissipation. This interaction ensures that the waves retain the imprint of their source, enabling astronomers to trace back the cosmic events that generated them.

The Harmony of Cosmic Waves

Cosmic waves, whether electromagnetic or gravitational, exemplify the dynamic interplay of stability and motion within the energy medium:

Universal Coherence: The energy medium maintains the harmony of wave propagation, ensuring that waves of different kinds—light waves, gravitational waves, and even quantum fluctuations—are interconnected within the same framework. This coherence is what allows the universe to function as a unified whole, with waves acting as the conduits of energy and information.

Interplay Between Waves and Structures: Cosmic waves interact with the material structures of the universe, such as stars, planets, and galaxies. These interactions often lead to fascinating phenomena, such as gravitational lensing, where light waves bend around massive objects due to distortions in the energy medium, or the absorption and re-emission of electromagnetic waves by interstellar matter.

Gravitational Waves: Evidence of Energy Dynamics

Gravitational waves provide a direct glimpse into the energy medium's dynamic nature. Their existence and behavior reinforce the idea that the universe is not an empty void but an interconnected system governed by energy dynamics:

Gravitational Waves as Proof of the Energy Medium: The propagation of gravitational waves through seemingly "empty" space highlights the presence of the stable energy pool. These waves could not exist or travel coherently without a medium to support their motion.

Energy Recycling Through Waves: Gravitational waves, like other cosmic phenomena, contribute to the continuous cycle of energy redistribution. The energy carried by these waves eventually dissipates into the stable energy pool, maintaining the universe's dynamic equilibrium.

Cosmic Waves: The Universal Symphony

The propagation of waves in the cosmos is more than just a mechanism of energy transfer; it is a universal symphony orchestrated by the stable energy pool. Whether it is the oscillations of electromagnetic fields or the ripples of gravitational waves, the energy medium ensures that these waves fulfill their roles in the grand cosmic dance.

By understanding cosmic wave dynamics through the lens of UED, we gain profound insights into how energy transitions sustain the harmony and evolution of the universe. This perspective not only unites diverse phenomena under a single framework but also highlights the central role of the energy medium as the unseen yet essential fabric of existence.

10.6 Frequency, Wavelength, and Energy Transfer

Waves, as manifestations of energy transitions, carry unique properties that reveal the intricacies of the universe's energy dynamics. Among these, frequency, wavelength, and energy transfer are interconnected in ways that transcend scales and media. These properties form the foundation of wave behavior, allowing energy to propagate and interact seamlessly within the universal energy medium.

The Connection Between Frequency, Wavelength, and Energy Transitions

The relationship between frequency, wavelength, and energy is fundamental to understanding wave mechanics. These properties are not isolated but are intricately linked through energy dynamics:

Frequency as a Measure of Energy Activity: Frequency represents the number of oscillations or cycles a wave undergoes in a given time. In electromagnetic waves, it reflects the vibratory motion of electric and magnetic fields. The higher the frequency, the greater the energy associated with the wave, as described by the equation E=hνE = h \nuE=hν, where EEE is energy, hhh is Planck's constant, and ν\nuν is frequency.

Wavelength and Its Role in Propagation: Wavelength, the distance between successive peaks or troughs of a wave, is inversely proportional to frequency (λ=c/ν\lambda = c / \nuλ=c/ν, where λ\lambdaλ is wavelength, ccc is the speed of light, and ν\nuν is frequency). This relationship underscores how energy distribution adapts to the medium and scale of propagation.

Energy Transitions and Wave Properties: Waves act as carriers of energy transitions. For example, photons in electromagnetic waves transport quantized energy packets that depend on their frequency. Similarly, mechanical waves transfer energy through vibrations, with their wavelength and frequency determined by the medium's properties.

Consistency Across Scales and Media

One of the most striking aspects of wave behavior is the consistency of their fundamental properties across varying scales and media. This universality is a direct reflection of the energy medium's role:

Frequency as a Universal Constant: Regardless of the medium, the frequency of a wave remains invariant during propagation. For electromagnetic waves, this constancy underscores the uniformity of energy transitions within the stable energy pool, ensuring that the energy carried by photons is not altered by changes in medium or scale.

Adaptability of Wavelength: While frequency remains constant, wavelength adapts to the medium, reflecting changes in the speed of wave propagation. For example, light waves slow down and experience shorter wavelengths when transitioning from air to water, but their frequency—and thus their energy—remains unchanged.

Energy Medium as the Unifying Factor: The stable energy pool provides a consistent framework that supports the seamless propagation of waves. It ensures that wave properties are preserved, even as waves encounter varying environments and scales. This interplay allows energy transitions to remain coherent across the universe.

Implications for Energy Transfer

The interplay between frequency, wavelength, and energy transfer has profound implications for understanding cosmic phenomena and the universal energy medium:

Energy Precision in Waves: The energy carried by a wave is precisely determined by its frequency, highlighting the fundamental role of vibratory motion in energy dynamics. This precision enables electromagnetic waves to transfer energy efficiently, whether in the form of visible light, radio signals, or gamma rays.

Universality Across Systems: From the quantum interactions of particles to the propagation of gravitational waves, the principles governing frequency, wavelength, and energy transfer remain consistent. This universality reflects the underlying harmony of the energy medium.

Bridging Micro and Macro Realms: The consistent behavior of wave properties across scales provides a bridge between the quantum and cosmic domains. It reveals that the same principles governing atomic-scale interactions also dictate the motion of celestial bodies and the behavior of interstellar waves.

Waves as Messengers of Energy Dynamics

In conclusion, the relationship between frequency, wavelength, and energy transitions encapsulates the essence of wave mechanics. Waves are more than just carriers of energy; they are dynamic expressions of the energy medium's stability and motion. By understanding how these properties interact and remain consistent across scales and media, we uncover the profound harmony of energy dynamics that governs the universe. Through the lens of Unified Energy Dynamics (UED), waves emerge not as isolated phenomena but as integral components of the universal interplay of energy.

10.7 Wave Behavior across Scales

Waves, as dynamic carriers of energy, exhibit behaviors that seamlessly bridge the quantum and macro scales. Whether propagating within atomic structures or traversing interstellar distances, their fundamental properties remain consistent, governed by the interplay of stable and active energy within the universal energy medium.

Wave Interactions at the Quantum Level

At the quantum scale, wave behavior is intricately tied to the foundational principles of energy transitions. Here, waves are not merely abstract concepts but tangible phenomena that shape the very fabric of matter and energy:

Wave-Particle Duality: Quantum entities like electrons and photons exhibit dual behavior, oscillating between particle-like and wave-like states. This duality reflects the vibratory motion of active energy within the stable energy medium, as described in the principles of Unified Energy Dynamics (UED).

Quantum Wave Functions: The behavior of quantum particles is governed by wave functions, mathematical descriptions that encapsulate probabilities of position and momentum. These wave functions underscore the energy medium's role in shaping quantum phenomena, with waves acting as expressions of localized disturbances.

Interactions at Subatomic Scales: Quantum waves interact with confined active energy, such as in particle collisions or during transitions between energy states. For example, the emission of photons during electron transitions in atoms illustrates the interplay between energy dynamics and wave behavior.

Macro-Scale Phenomena: Radiation and Interstellar Signals

At larger scales, waves manifest in phenomena that define the cosmos, from electromagnetic radiation to gravitational ripples. These macro-scale interactions highlight the universal applicability of wave principles:

Electromagnetic Radiation: Light waves, radio signals, and gamma rays propagate through the energy medium, carrying energy across vast cosmic distances. Their behavior, characterized by frequency and wavelength, reflects the energy dynamics of their origin, whether a distant star or a human-made transmitter.

Gravitational Waves: Generated by massive cosmic events, such as the merging of black holes, gravitational waves represent ripples in the stable energy pool. These waves demonstrate the ability of the energy medium to sustain and transmit disturbances across unimaginable distances, providing insights into the universe's structure and motion.

Interstellar Communication: Waves are vital for transmitting information across the cosmos. From natural signals like pulsar emissions to artificial communications sent by probes, wave behavior enables humanity to explore and connect with the broader universe. The consistency of wave properties, such as frequency and energy transfer, ensures the reliability of these transmissions.

Bridging Quantum and Cosmic Realms

The behavior of waves across scales reveals the interconnectedness of the quantum and macro domains. Despite their vastly different contexts, both scales adhere to the same principles of wave mechanics, emphasizing the energy medium's unifying role:

Universal Consistency: The frequency of waves remains invariant, whether in subatomic interactions or interstellar propagation. This consistency highlights the stability of the energy medium as a framework for energy transitions.

Dynamic Adaptation: While frequency remains constant, wavelength and amplitude adapt to the medium and environment, enabling waves to transfer energy effectively across diverse conditions.

Harmonized Functionality: Waves serve as the bridge between micro and macro realms, translating energy transitions into observable phenomena that resonate with the underlying harmony of the cosmos.

Waves as Universal Messengers

In short, the behavior of waves, from quantum oscillations to cosmic signals, demonstrates their integral role in energy dynamics. By understanding how waves interact across scales, we uncover the cohesive framework that connects the smallest particles to the largest cosmic structures. Unified Energy Dynamics (UED) offers a lens through which wave behavior is not fragmented but harmoniously unified, reflecting the timeless principles of energy that govern the universe. Waves, in their vibratory dance, embody the universal language of energy transitions, sustaining the rhythm of existence itself.

10.8 The Infinite Energy Medium as a Wave Conductor

The infinite energy medium, as described in Unified Energy Dynamics (UED), is not merely a backdrop but an active participant in the propagation and behavior of waves. Serving as the universal conductor, it ensures that waves—whether electromagnetic, gravitational, or otherwise—maintain coherence and harmony as they traverse the vast expanse of the cosmos. By understanding this role, we gain profound insights into the interaction of waves with matter and the energy pool itself.

Ensuring Harmony and Coherence in Wave Dynamics

The energy medium is the steady framework that supports the propagation of waves, maintaining their integrity and facilitating their interactions across different scales and environments:

Wave Propagation Without Material Medium: Unlike mechanical waves, which rely on the vibration of particles within a physical medium, electromagnetic waves move through the energy medium without requiring matter. The oscillating electric and magnetic fields of these waves interact directly with the stable energy pool, enabling seamless transmission even in the vacuum of space.

Coherence Across Cosmic Scales: The energy medium ensures that waves remain coherent, preserving their frequency and phase relationships over vast distances. For example, light from distant stars retains its spectral characteristics despite traversing billions of light-years, demonstrating the medium's role in maintaining wave integrity.

Harmonized Transitions: As waves encounter different environments—such as interstellar gas, planetary atmospheres, or dense gravitational fields—the energy medium facilitates smooth transitions. It allows waves to adapt their properties, such as wavelength and amplitude, while preserving their fundamental energy dynamics.

Wave-Matter Interaction Through UED

The interaction of waves with matter reveals the intricate connection between the energy medium and physical reality. These interactions are not isolated phenomena but expressions of the universal principles of energy dynamics:

Energy Transfer Mechanisms: When waves interact with matter, they facilitate energy transfer at various levels.

For instance:

Photoelectric Effect: Photons impart their energy to electrons, demonstrating the seamless exchange of active energy between waves and matter.

Gravitational Waves: Ripples in the energy medium caused by massive cosmic events can subtly influence matter, revealing the interplay between macroscopic disturbances and the medium's stability.

Wave Absorption and Emission: Matter can absorb and emit waves as part of energy transitions, exemplifying the energy medium's role as a universal conductor.

For example:

Emission of Photons: Electrons transitioning between energy states release photons, a direct interaction with the energy medium.

Absorption of Radiation: Matter absorbs energy from waves, leading to phenomena like heating or chemical reactions.

Reflection, Refraction, and Diffraction: These classic wave behaviors are manifestations of how waves navigate and adapt to changes in their environment. The energy medium provides the framework for these interactions, ensuring that energy is redistributed in a coherent and predictable manner.

Insights into Universal Wave Conductivity

The energy medium's role as a conductor transcends traditional concepts of wave mechanics, offering a unified perspective on wave dynamics:

Universal Connectivity: The energy medium bridges the micro and macro realms, enabling waves to propagate and interact seamlessly across scales. This connectivity underscores the unity of physical and metaphysical phenomena within the UED framework.

Stability and Dynamism: While waves represent dynamic energy transitions, the medium's stable nature ensures that these transitions occur within a coherent and balanced system. This duality is the hallmark of UED, reflecting the harmonious interplay of stability and motion.

A Framework for Cosmic Exploration: By understanding wave dynamics through the lens of the energy medium, we gain new tools for exploring the universe. From studying distant galaxies through their electromagnetic emissions to detecting gravitational waves, the energy medium provides the key to unlocking cosmic mysteries.

Waves in the Energy Medium

Thus, the infinite energy medium, as the universal wave conductor, ensures that waves remain coherent and functional across the vast expanse of the cosmos. Its role extends beyond facilitating propagation—it governs the very principles of wave-matter interaction, maintaining harmony in the dynamic dance of energy transitions. Through UED, we see waves not as isolated phenomena but as integral components of a unified and interconnected universe, perpetually shaped and sustained by the energy medium.

10.9 Conclusion: Waves as Universal Connectors

Waves, in their myriad forms, serve as the universal carriers of energy, bridging the vast expanse of the cosmos and connecting the micro and macro realms of existence. They are not just mechanisms of energy transfer but also expressions of the intricate interplay between stability and dynamism that underpins the universe. Through Unified Energy Dynamics (UED), waves emerge as vital components of the universal fabric, harmonizing energy transitions and interactions.

UED reframes our understanding of wave mechanics, uniting the principles that govern electromagnetic, gravitational, and quantum waves under a cohesive framework. By recognizing the energy medium as the fundamental conduit for wave propagation, UED reveals the harmony and coherence that sustain the universe's dynamic equilibrium. Waves are no longer isolated phenomena; they are universal connectors, perpetually shaping and reshaping the cosmos.

In this light, waves become more than just carriers of energy—they are manifestations of the universe's inherent unity and order. Through their continuous interaction with the infinite energy medium, they maintain the balance and flow that define existence. UED invites us to see waves not just as tools of exploration but as profound symbols of the interconnectedness of all things, weaving the cosmos into a seamless and dynamic whole.

Chapter 11: Cosmic Harmony: Unified Energy in Action

The universe is not a collection of isolated phenomena but a grand, interconnected system governed by the principles of energy dynamics. From the smallest quark to the largest galaxy, every particle and structure participates in a cosmic symphony of creation, transformation, and dissolution. Unified Energy Dynamics (UED) offers a lens through which to view this intricate interplay, revealing a harmony that transcends traditional distinctions between the micro and macro realms.

In the previous chapters, we explored the foundational concepts of the energy medium, the mechanics of waves, and the role of photons as universal messengers. These discussions set the stage for understanding how energy interactions shape the universe across scales, creating a cohesive and balanced cosmos. This chapter delves deeper into the unified principles that govern these interactions, highlighting the elegance and coherence of the universal dance driven by energy's stability and dynamism.

11.1 Introduction: The Universal Symphony of Energy

The universe operates as a vast symphony, where every particle and cosmic structure plays a role in a harmonious interplay of forces and dynamics. This grand orchestration, from the smallest quantum fluctuations to the motion of galaxies, is driven by energy transitions that uphold balance and coherence across scales. Unified Energy Dynamics (UED) provides a framework to understand this cosmic harmony, bridging the micro and macro realms of existence.

Through UED, we see that the same principles guiding the behavior of quarks also govern the movements of celestial bodies. The stable energy pool serves as the universal stage, enabling interactions that create, transform, and sustain all matter and phenomena. This chapter embarks on a journey to explore how energy dynamics unify the universe, revealing the interconnectedness and elegance that define its very nature.

11.2 Energy Interactions across Scales

The universe, in all its complexity, operates on principles that transcend scale, linking the behavior of the smallest particles to the vastest celestial structures. This continuum of energy dynamics is a cornerstone of Unified Energy Dynamics (UED), offering a framework that reveals the inherent unity of the cosmos. From quarks vibrating within atomic nuclei to galaxies swirling in cosmic dances, the same energy interactions govern the fundamental processes of creation, transformation, and dissolution.

Quarks and Subatomic Dynamics

At the quantum level, energy manifests as confined active energy within particles like quarks. These fundamental constituents of matter exhibit dynamic behaviors such as:

Wave-Particle Duality: Quarks, like photons, display both particle-like and wave-like properties, underscoring the role of vibratory motion as the signature of active energy.

Color Charge Interactions: Within the nucleus, quarks exchange gluons, binding them together in a dynamic equilibrium. This process reflects the energy transitions inherent in UED, where stability arises from active interactions within the stable energy pool.

Quantum Entanglement: Quarks can exhibit entangled states, where their properties remain interconnected despite spatial separation. This behavior mirrors the universal coherence sustained by the energy medium.

These subatomic dynamics illustrate how energy interactions are not random but guided by principles that ensure both stability and dynamism.

Atomic and Molecular Realms

Moving up the scale, the behavior of atoms and molecules continues to reflect energy dynamics:

Electron Transitions: Electrons orbiting nuclei transition between energy levels, emitting or absorbing photons in the process. These transitions are evidence of active energy interactions within the stable energy medium.

Chemical Reactions: The making and breaking of chemical bonds are driven by energy redistribution, with stable energy acting as the reservoir that balances these transformations.

The interplay of energy within atoms and molecules lays the foundation for the complexity of matter, including biological systems and inorganic structures.

Celestial Bodies and Galactic Motion

At the macro scale, energy dynamics govern the behavior of celestial bodies and their interactions:

Gravitational Interactions: As discussed in UED, gravity emerges as the interaction between mass and the stable energy pool. This interaction shapes the orbits of planets, the formation of stars, and the clustering of galaxies.

Stellar Evolution: Stars convert confined active energy in their cores into radiant energy, fueling their lifecycles. This process, which begins with the gravitational collapse of gas clouds, exemplifies how energy transitions drive cosmic phenomena.

Galactic Dynamics: The motion of galaxies, often attributed to the influence of dark matter, can be reconsidered through UED as disturbances in the stable energy medium, propagating balance and coherence across cosmic scales.

Unified Principles Across Scales

The beauty of UED lies in its ability to unify these seemingly disparate phenomena under a single framework:

Energy as the Foundation: Whether in the vibratory motion of quarks or the gravitational pull of galaxies, energy is the common thread that connects all interactions.

Dynamic Equilibrium: Stability at every scale is maintained by the balance between active and stable energy, ensuring that the cosmos remains coherent and harmonious.

Continuity and Interconnectedness: The principles governing micro and macro realms are not isolated but part of a seamless continuum, reflecting the infinite nature of the energy pool.

Implications for Understanding the Cosmos

Recognizing energy interactions across scales reshapes our perspective on the universe:

✓ It highlights the elegance of a system where the same principles govern the smallest and largest phenomena.

✓ It offers a pathway to reconcile quantum mechanics and relativity, traditionally seen as incompatible, by emphasizing their shared foundation in energy dynamics.

✓ It reveals the cosmos as an interconnected whole, where every action and interaction contributes to a grand, unified tapestry.

Through UED, we transcend the limitations of fragmented scientific disciplines, embracing a vision of the universe that is as cohesive as it is profound.

11.3 Gravitational Waves: Ripples in the Energy Pool

Gravitational waves, first predicted by Einstein's General Theory of Relativity and later confirmed through direct observation, represent a profound cosmic phenomenon. These waves are ripples in spacetime, traditionally understood to be caused by the acceleration of massive objects, such as merging black holes or neutron stars. Through the lens of Unified Energy Dynamics (UED), gravitational waves can be reimagined as ripples within the stable energy medium itself, carrying energy across vast distances while maintaining the harmony of the cosmos.

Gravitational Waves in UED

In the framework of UED, gravitational waves are not distortions in spacetime per se but disturbances propagated through the infinite stable energy pool. These waves are the physical manifestations of energy redistributing within the medium, maintaining equilibrium in response to massive cosmic events:

Ripples of Energy: When massive objects interact, such as in the collision of black holes, they create localized disturbances in the energy medium. These disturbances propagate outward as gravitational waves, much like ripples on the surface of a pond.

Energy Redistribution: The emission of gravitational waves represents a mechanism for transferring active energy through the stable medium, ensuring that the equilibrium of the universal energy pool is preserved.

Propagation Through the Stable Energy Medium

Gravitational waves are unique among cosmic phenomena in that they interact minimally with matter, allowing them to traverse vast cosmic distances largely undisturbed:

Medium-Dependent Propagation: While traditional waves require a physical medium for propagation, gravitational waves leverage the inherent stability of the energy pool. The energy medium serves as the substrate that sustains and propagates these waves, enabling their transmission without the need for material interaction.

Frequency and Amplitude: The characteristics of gravitational waves—such as frequency and amplitude—are directly tied to the nature of the energy transitions that create them. Higher-energy events produce stronger, shorter-wavelength waves, while less energetic events result in weaker, longer-wavelength ripples.

Insights from Gravitational Waves

Gravitational waves provide a unique window into the universe, revealing information about events that might otherwise be inaccessible:

Cosmic Diagnostics: By studying gravitational waves, we gain insights into the dynamics of massive objects, such as black hole mergers, supernovae, and even the early universe. These waves carry the "signature" of the energy interactions that produced them, offering direct evidence of the interplay between active and stable energy.

Energy Medium Validation: The ability of gravitational waves to propagate through the energy medium without significant attenuation supports the idea of an infinite, stable substrate that unifies all energy interactions.

Revisiting Gravitational Waves Through UED

The UED perspective enhances our understanding of gravitational waves by shifting the focus from spacetime distortions to energy medium dynamics:

Unifying Framework: Gravitational waves are no longer seen as an isolated relativistic phenomenon but as an integral part of the broader energy dynamics governing the cosmos.

Energy Continuity: These waves exemplify the seamless flow of active energy within the stable energy medium, reinforcing the interconnectedness of all cosmic processes.

By interpreting gravitational waves as ripples in the stable energy pool, UED provides a more comprehensive understanding of their origin, propagation, and significance. This perspective not only aligns with observed phenomena but also deepens our appreciation of the universe's intricate and harmonious design.

11.4 Cosmic Structures and Energy Stability

The universe, in its vast complexity, is a masterpiece of energy interactions. From the formation of galaxies to the intricate dynamics of planetary systems, cosmic structures are governed by the delicate interplay between stability and motion. Unified Energy Dynamics (UED) reveals the fundamental role energy stability plays in shaping these colossal entities, offering insights into how the universe maintains its coherence while fostering dynamic evolution.

Energy Dynamics in Galactic Formation

Galaxies, the fundamental building blocks of the cosmos, are born from the intricate dance of energy stability and motion:

Energy Disturbances and Galaxy Formation: Stable energy disruptions in the universal medium lead to the condensation of matter, forming stars and planets within galaxies. These disruptions create localized pockets of active energy that evolve into dynamic systems.

Gravitational Stability: The gravitational interaction between mass concentrations, mediated by the energy medium, ensures that galaxies maintain their structural coherence. This balance allows galaxies to retain their spiral or elliptical shapes despite the chaotic motion of their constituent stars and gas clouds.

Stars and the Interplay of Stability and Motion

Stars are dynamic hubs of energy transformation, showcasing the profound role of active energy within cosmic structures:

Energy Concentration: Stars emerge when regions of stable energy in the universal medium transition into active energy, initiating nuclear fusion. This transformation fuels the star's lifecycle and radiates energy across vast distances.

Dynamic Equilibrium: The internal pressure from nuclear reactions balances the gravitational pull, allowing stars to maintain their size and shape. This equilibrium reflects the cosmic harmony between stability and dynamism.

Energy Redistribution: As stars evolve, they redistribute energy through radiation and stellar winds, contributing to the formation of new structures such as planets and nebulae.

Planetary Systems: Microcosms of Cosmic Dynamics

Planetary systems, like our solar system, are microcosms of the larger universal energy dynamics:

Formation from Accretion: Planets form from the accretion of matter within protoplanetary disks, regions where stable energy condenses into active forms. This process showcases the cyclical nature of energy transitions, as active energy manifests in diverse planetary forms.

Orbital Harmony: The stable energy medium ensures the synchronization of planetary orbits around stars. Gravitational interactions, guided by the medium, prevent chaotic motion and foster long-term stability in these systems.

Energy Exchange: Planetary systems participate in energy redistribution through phenomena such as tidal interactions, atmospheric dynamics, and the capture or release of meteoroids and comets.

Large-Scale Cosmic Architecture

The interplay between stability and motion governs the structure and evolution of the universe itself:

Cosmic Web Formation: The universe's large-scale structure, the cosmic web, emerges from the stable energy medium's ability to support matter distribution and gravitational interactions across vast scales. Filaments of galaxies and clusters are the result of this delicate energy balance.

Dynamic Stability: The stability of the universal energy pool underpins the expansion of the universe while allowing localized motion and transformation. This dynamic stability ensures that galaxies evolve without disintegration and that new structures continuously emerge.

Cosmic Harmony Through Energy Stability

The intricate balance of stability and motion reveals the universe as a harmonious interplay of energy dynamics:

Stability as the Framework: Stable energy provides the foundation for the coherence of cosmic structures, ensuring that the vast tapestry of galaxies, stars, and planets operates within an orderly framework.

Motion as the Catalyst: Active energy introduces the dynamism necessary for evolution, creating the motion that drives transformations and sustains the universe's vibrancy.

Reframing Cosmic Structures Through UED

Unified Energy Dynamics offers a holistic perspective on cosmic architecture:

Integration Across Scales: By linking the micro and macro realms, UED unifies the principles governing particles and galaxies, revealing a continuous flow of energy that sustains cosmic structures.

Harmony in Complexity: The universe's complexity is not chaotic but a reflection of the seamless interplay between stability and motion, guided by the infinite energy medium.

Through the lens of UED, cosmic structures emerge not as isolated phenomena but as interconnected expressions of universal energy. This understanding deepens our appreciation of the cosmos's beauty, coherence, and infinite potential.

11.5 The Interconnection of Forces and Energy Dynamics

The forces of nature—gravitational, electromagnetic, and nuclear—are often studied in isolation within their respective domains. However, Unified Energy Dynamics (UED) reveals that these forces are not independent phenomena but interconnected expressions of energy transitions within the infinite energy medium. Through UED, we can understand how these forces work in harmony to sustain the cosmos's intricate balance and dynamic evolution.

Gravitational Force: The Framework of Cosmic Order

Gravity as an Energy Interaction: UED redefines gravity not as a standalone force but as the interaction between mass and the stable energy medium. Massive objects disturb the energy medium, creating an effect that guides the motion of other masses.

Cosmic Harmony Through Gravity: Gravity ensures the stability of large-scale structures, from planetary systems to galaxies, by balancing motion and mass concentration. This interplay maintains the coherence of celestial mechanics and the universe's architectural order.

Dynamic Role in Energy Flow: Gravity also facilitates the reintegration of active energy into the stable energy pool, as seen in phenomena like black hole formation. This cyclical process underscores gravity's role in the universal flow of energy.

Electromagnetic Force: The Driver of Interaction

Fields in Oscillation: Electromagnetic forces emerge from the dynamic interplay of electric and magnetic fields within the energy medium. These fields enable the transfer and transformation of energy across vast distances.

Universal Applicability: Electromagnetic forces govern a wide range of phenomena, from the behavior of photons to chemical bonding. They mediate interactions at the atomic level and propagate energy through light and other electromagnetic waves.

Connecting Micro and Macro Realms: Electromagnetic forces bridge the micro and macro scales, influencing processes as diverse as quantum transitions in atoms and the radiation emitted by stars.

Nuclear Forces: The Stabilizers of Matter

Strong Nuclear Force: The strong nuclear force binds protons and neutrons within atomic nuclei, overcoming the repulsive electromagnetic force between positively charged protons. This force ensures the stability of matter at the quantum level.

Weak Nuclear Force: The weak nuclear force governs processes like radioactive decay and particle interactions, playing a crucial role in energy release during nuclear reactions.

Energy Transitions in Nuclear Reactions: Nuclear forces facilitate energy transitions by converting active energy stored in nuclei into other forms, such as photons or kinetic energy. These reactions are pivotal in stars and other high-energy environments.

Harmony Among Forces

Unified Principles: UED shows that all forces are facets of energy dynamics, arising from the interactions of stable and active energy within the universal medium. This perspective eliminates the apparent separation between forces and reveals their shared foundation.

Complementary Roles: Gravitational, electromagnetic, and nuclear forces work in tandem to sustain the universe's complexity:

✓ Gravity governs large-scale structures, ensuring stability and coherence.

✓ Electromagnetic forces drive interaction and energy transfer across scales.

✓ Nuclear forces stabilize matter, enabling the formation of elements and the release of energy in stellar processes.

Synergy Across Scales: These forces interact seamlessly, creating a harmonious balance that supports both the smallest particles and the largest galaxies.

Unified Energy Dynamics as the Underlying Thread

Energy as the Common Origin: UED identifies energy as the source from which all forces emerge, highlighting their interconnected nature. The infinite energy medium acts as the stage on which these forces operate, ensuring their coordination and balance.

Dynamic Interplay: The forces' interactions reflect the constant exchange between stability and motion, active and stable energy, within the universal framework. This dynamic interplay is the essence of the cosmos's unity and evolution.

Guiding Universal Processes: By understanding forces through UED, we gain insights into the mechanisms that govern cosmic harmony, from the behavior of particles to the evolution of galaxies.

Reframing Forces Through UED

Through the lens of UED, the forces of nature are no longer isolated phenomena but interconnected expressions of energy. Gravitational, electromagnetic, and nuclear forces are the agents of energy dynamics, shaping the universe's structure, behavior, and evolution. This unified perspective not only enhances our understanding of the cosmos but also underscores the profound harmony underlying its complexity.

11.6 Cosmic Evolution: A Harmonious Dance

The cosmos is a symphony of energy transitions, where the creation, transformation, and dissolution of matter unfold in a seamless, cyclical process. Unified Energy Dynamics (UED) offers a profound lens through which to understand this intricate dance, revealing how the interplay of stability and motion shapes the universe's evolution at every scale.

Energy Transitions as Drivers of Cosmic Change

Creation of Matter: Matter emerges from disturbances in the stable energy pool, where localized transitions of energy give rise to particles and structures. This initial formation sets the stage for the development of stars, planets, and galaxies.

Transformation of Systems: Once formed, matter undergoes continual transformation driven by energy transitions. Processes such as nuclear fusion in stars, chemical reactions on planetary surfaces, and gravitational interactions across celestial bodies exemplify the dynamic evolution of matter and energy.

Dissolution and Reintegration: Just as energy transitions create matter, they also facilitate its dissolution. When matter collapses—such as during supernovae or black hole formation—it returns active energy to the stable energy pool, completing the cycle of existence.

The Cyclical Nature of Cosmic Evolution

The Eternal Cycle: UED emphasizes that the universe is not a linear progression but a perpetual cycle of creation, transformation, and reintegration. This cycle maintains balance and harmony within the cosmos.

✓ **Birth:** Stars ignite as energy transitions within clouds of gas and dust.

✓ **Life:** Galaxies evolve as matter interacts under gravitational and electromagnetic forces.

✓ **Death:** Stellar remnants release energy, enriching the universe and returning to the stable pool.

Universal Continuity: The cyclical nature of cosmic evolution ensures that no energy is ever lost. Instead, it transforms and redistributes, maintaining the dynamic equilibrium of the infinite energy medium.

Harmony through UED Principles

Interplay of Stability and Motion: Cosmic evolution reflects the delicate balance between stable and active energy. Stability provides the framework for coherence, while motion drives change and diversity.

Unified Dynamics Across Scales: From the quantum fluctuations that birth particles to the gravitational waves propagating through galaxies, the same UED principles govern the behavior of the universe at all levels.

Cosmic Purpose: This harmonious dance reveals an intrinsic order, where every event, from the smallest particle interaction to the largest galactic merger, contributes to the universe's perpetual flow of energy.

A Vision of Eternal Balance

Cosmic evolution, as seen through UED, is not random but deeply orchestrated. The universe thrives on the constant interplay of forces and energy states, where every transition contributes to a greater equilibrium. By embracing the cyclical nature of creation and dissolution, we glimpse the profound unity and order that govern the cosmos—a harmonious dance sustained by the timeless principles of energy dynamics.

11.7 Stability and Dynamism in Universal Structures

The universe thrives on a delicate balance between stability and dynamism, two complementary principles that govern its structure and evolution. Stability provides the coherence needed to maintain order, while dynamism fuels the transformation and growth that shape cosmic phenomena. Unified Energy Dynamics (UED) illuminates this interplay, offering insights into how these forces sustain the harmonious architecture of the cosmos.

Stability: The Framework of Coherence

Foundation of Universal Order: Stability in the energy medium acts as the silent backbone of existence, ensuring that the laws of physics remain consistent across time and space. This stability enables the formation and persistence of celestial bodies, such as stars, planets, and galaxies.

Energy Pool as the Stable Reservoir: The stable energy pool absorbs disturbances and sustains equilibrium, allowing cosmic structures to exist within a coherent framework.

For instance:

Planetary Orbits: Stability in the energy medium ensures the predictable motion of planets around their stars, maintaining the delicate balance required for life to emerge.

Galactic Structures: Stability allows galaxies to retain their spiral or elliptical shapes over billions of years, preserving their architectural coherence amidst the chaos of interstellar dynamics.

Dynamism: The Catalyst for Transformation

Driving Cosmic Change: Dynamism arises from the interplay of active energy, which initiates motion, interaction, and transformation within the stable energy pool.

It is the force behind:

✓ **Star Formation:** Gravitational collapse within gas clouds transforms potential energy into active energy, igniting nuclear fusion and birthing stars.

✓ **Supernovae:** The explosive death of massive stars releases vast amounts of active energy, scattering heavy elements across space and enriching future stellar systems.

Continuous Evolution: Dynamism ensures that the universe remains vibrant and ever-changing. Without it, cosmic structures would stagnate, lacking the forces necessary for growth and adaptation.

Interplay Between Stability and Dynamism

The interaction between stability and dynamism creates the dynamic equilibrium that defines the cosmos. Neither principle dominates; instead, they work in concert to sustain universal harmony.

Cosmic Phenomena Exemplifying This Interplay:

- ✓ **Black Holes:** Stability ensures the coherence of spacetime around black holes, while dynamism governs the immense energy release through radiation and matter accretion.

- ✓ **Gravitational Waves:** These ripples in the energy medium represent dynamism, propagating through the stable framework of space to convey energy transitions across vast distances.

- ✓ **Galactic Collisions:** While seemingly chaotic, galactic mergers illustrate the balance of stability and dynamism. The galaxies' stable cores merge to form new structures, while dynamic interactions reshape their stars and gas clouds.

Balancing Coherence and Transformation

Micro to Macro Scale: At the quantum level, stability ensures the persistence of particles, while dynamism drives interactions like entanglement and wave-particle duality. On the macro scale, stability anchors celestial bodies, while dynamism catalyzes processes like supernovae and planetary formation.

Sustaining Cosmic Harmony: The universe's ability to maintain coherence while embracing transformation reflects the unified principles of UED. Every event, from the smallest atomic vibration to the largest galactic collision, contributes to this balance.

A Universe in Dynamic Equilibrium

The interplay of stability and dynamism is the essence of cosmic harmony. Stability offers a steady framework, enabling the universe to persist and function cohesively. Dynamism, on the other hand, injects vitality into this framework, driving the cycles of creation, transformation, and dissolution. Together, they create a cosmos that is both ordered and alive—a masterpiece of perpetual balance governed by the timeless principles of energy dynamics.

11.8 The Role of Energy Pool in Sustaining Harmony

The infinite energy pool serves as the cornerstone of universal harmony, providing the stable foundation upon which all cosmic phenomena are built. As the unifying medium of existence, the energy pool ensures coherence, continuity, and balance across scales, from the quantum to the galactic. Unified Energy Dynamics (UED) highlights its pivotal role in sustaining the intricate interplay of forces and matter that defines the universe.

The Infinite Energy Medium: A Universal Constant

Source of Stability: The energy pool acts as an infinite reservoir of stable energy, absorbing disturbances and maintaining equilibrium. Its omnipresence ensures that even in the face of dynamic transformations, the fundamental order of the cosmos remains intact.

Dynamic Yet Coherent: While the energy medium accommodates the constant motion of active energy, it stabilizes these interactions, preventing chaos and preserving the underlying structure of the universe.

For instance, gravitational waves ripple through this stable pool without disrupting its foundational equilibrium. Similarly, electromagnetic waves propagate seamlessly, demonstrating the harmony inherent in the energy medium.

Equilibrium Across Cosmic Scales

Quantum Stability: At the smallest scales, the energy pool ensures that particles maintain coherence even during interactions like entanglement or annihilation. This stability enables the persistence of matter and the emergence of complex structures.

Celestial Balance: On a larger scale, the energy pool sustains the delicate equilibrium required for celestial phenomena.

✓ **Planetary Systems:** The stable energy medium provides the framework for planets to orbit their stars in synchronized motion, ensuring the predictability of cosmic mechanics.

✓ **Galactic Formation:** The energy pool absorbs and redistributes the forces generated during galactic mergers or star formation, preventing catastrophic imbalances.

Gravitational Harmony: Gravity, as an interaction within the energy medium, operates in sync with other forces to maintain cosmic order. The energy pool acts as the mediator, ensuring that gravitational interactions neither destabilize the system nor disrupt the cohesion of the universe.

Insights into Universal Balance

Cyclical Energy Flow: The energy medium facilitates the cycle of energy emergence, transformation, and reintegration. This continuous flow ensures that no energy is lost, maintaining the universe's dynamic equilibrium.

Unified Principles: The energy pool's role as the ultimate unifier allows disparate phenomena—like the behavior of photons, the propagation of waves, and the dynamics of black holes—to coexist within a single coherent framework.

Cosmic Resilience: By absorbing disturbances and redistributing energy, the energy pool ensures that the universe remains resilient against disruption, whether at the scale of a particle collision or a galactic supernova.

Sustaining Harmony Through UED

Unified Energy Dynamics reveals the infinite energy pool as more than a passive backdrop—it is the active foundation of the universe's balance and coherence. By connecting all phenomena through stable and active energy interactions, it ensures that the cosmos functions as a harmonious whole. Whether shaping quantum behavior, guiding planetary systems, or sustaining the rhythm of gravitational waves, the energy pool is the silent architect of universal harmony.

11.9 Conclusion: The Universe as a Unified Whole

The cosmos, in all its grandeur and complexity, is far from a chaotic assembly of isolated phenomena. It is an intricately interconnected whole, bound together by the universal principles of energy dynamics. Unified Energy Dynamics (UED) offers a transformative lens through which this interconnectedness becomes not only apparent but deeply meaningful. By revealing the relationships between stability and dynamism, matter and energy, and the micro and macro realms, UED uncovers the harmony underlying the universe's infinite complexity.

Reaffirming the Interconnectedness of Cosmic Phenomena

From Particles to Galaxies: UED demonstrates that the same principles governing the behavior of subatomic particles also dictate the formation and evolution of galaxies. This continuity across scales is evidence of the universe's inherent unity.

✓ **Quantum Scale:** The vibratory motion of particles, the emergence of photons, and the interplay of electromagnetic fields are all manifestations of energy transitions.

✓ **Cosmic Scale:** Gravitational waves, galactic rotations, and the birth and death of stars reflect the same energy principles, magnified across vast distances.

The Energy Pool as the Unifying Medium: The infinite energy pool is the silent thread connecting all phenomena. Whether stabilizing quantum interactions or sustaining the gravitational balance of celestial systems, the energy pool ensures that all processes operate within a coherent framework.

Harmony in Diversity

Balance Between Stability and Dynamism: The universe's complexity arises from the interplay of stability and motion. Stable energy provides the foundation upon which active energy manifests as transformation and creation.

✓ **Dynamic Equilibrium:** The rhythmic cycle of energy emergence, transformation, and reintegration ensures that the universe remains dynamic yet balanced.

✓ **Cosmic Symphony:** Each disturbance in the energy medium, whether the formation of a star or the propagation of gravitational waves, contributes to the harmonious whole.

Integration of Forces: UED reveals that gravitational, electromagnetic, and nuclear forces are not disparate entities but facets of a single energy principle. Their interplay sustains the cosmic structure, from the quantum realm to the largest galactic formations.

Revealing Harmony Through Complexity

Order in Chaos: What may appear as chaos—like the unpredictable behavior of quantum particles or the turbulent interactions of galaxies—is, through UED, revealed as an expression of underlying order. The energy pool ensures that every motion and transformation aligns with universal principles.

Cycles of Creation and Dissolution: The cyclical nature of the universe, where energy transitions continuously between stable and active states, embodies a profound symmetry. From the emergence of matter to its dissolution into the energy pool, this cycle reflects a perpetual dance of harmony.

Implications of UED for Understanding the Universe

A Unified Perspective: By integrating scientific observation, philosophical insight, and metaphysical understanding, UED transcends traditional boundaries. It provides a cohesive framework for exploring the universe's most profound mysteries.

New Pathways for Exploration: UED not only refines our understanding of existing phenomena but also opens the door to novel inquiries. It challenges us to reconsider established assumptions, such as the nature of time, space, and gravity, and inspires innovative approaches to cosmic exploration.

A Harmonious Cosmos

The universe is not a fragmented collection of phenomena but a unified whole, where every particle, wave, and celestial body is intricately connected. UED reveals this interconnectedness by emphasizing the foundational role of energy in shaping all aspects of existence. From the silent stability of the energy pool to the vibrant dynamism of active energy, the cosmos is a testament to balance, coherence, and infinite potential.

By embracing UED, we gain not only a deeper understanding of the universe but also a renewed sense of wonder for its harmonious complexity. It is a reminder that amidst the vastness and mystery of the cosmos, there is an underlying unity—a universal rhythm that binds us all to the infinite dance of energy.

Chapter 12: Atomic and Quantum Phenomena in Energy Dynamics

As we delve deeper into the microcosm of the universe, we encounter a realm where the principles of Unified Energy Dynamics (UED) manifest in their most intricate and fascinating forms. At the atomic and quantum levels, energy interactions drive the very foundation of matter, shaping the building blocks of existence. From the vibratory motion of particles to the mysterious behavior of photons and electrons, this chapter explores how UED provides a cohesive understanding of phenomena that have long puzzled scientists and philosophers alike.

Transitioning from the cosmic harmony of galaxies and gravitational waves, we now shift our focus to the subatomic stage, where stability and dynamism create a delicate balance. Here, particles emerge as localized manifestations of energy disturbances, their behavior dictated by the intricate interplay of the stable energy pool and active energy states. Through this lens, UED not only bridges the quantum and classical worlds but also reveals a profound unity underlying all scales of existence.

12.1 The Fundamental Role of Atomic Particles

At the heart of matter lies a profound interplay of energy, embodied by the atomic particles that form its structure. Protons, neutrons, and electrons—often seen as the building blocks of atoms—are far more than mere physical entities. Within the framework of Unified Energy Dynamics (UED), these particles are manifestations of active energy, intricately interacting with the stable energy pool to give rise to the phenomena we observe in the material world.

Protons and Neutrons: Anchors of Stability

Protons and neutrons reside in the nucleus, forming the core of the atom. These particles are concentrated expressions of active energy, bound tightly together by the strong nuclear force. Their presence ensures the atom's structural integrity and neutrality:

Protons carry a positive charge, creating electromagnetic interactions that extend beyond the nucleus. This charge emerges from energy imbalances within the stable energy medium, marking the proton as a key player in energy transitions.

Neutrons are charge-neutral yet indispensable. They balance the nucleus by mitigating the repulsive forces between protons, while their latent energy participates in nuclear reactions, such as decay and fusion.

Electrons: Dynamic Messengers of Energy

Electrons, existing in the extranuclear region, represent the dynamic and transformative nature of active energy. Their motion and interactions drive countless phenomena:

Orbital Motion: Electrons move in quantized energy levels around the nucleus, their behavior dictated by the energy medium's influence. This motion encapsulates the duality of stability (orbitals) and dynamism (transitions between energy states).

Energy Transitions: When electrons absorb or release energy, they transition between orbits, emitting photons or other particles. These transitions reflect the ongoing dialogue between active energy and the stable energy pool.

The Emergence of Charge and Fields

The presence of protons, neutrons, and electrons gives rise to electric and magnetic fields, fundamental properties of matter:

Charge as Energy Imbalance: Positive and negative charges emerge from localized disturbances in the energy medium, ensuring the potential for interaction and transformation.

Fields as Extensions of Energy Influence: Electric and magnetic fields are the manifestations of these charges' interactions with the energy medium. They enable the propagation of forces and energy transitions, acting as the "wings" that carry active energy across systems.

Atomic Particles as Bridges Between Micro and Macro

Through their roles, atomic particles bridge the micro and macro realms. Their behavior at quantum scales governs chemical reactions, material properties, and even large-scale structures like stars and galaxies. Within the UED framework, protons, neutrons, and electrons become the conduits through which energy manifests, transforms, and sustains the physical reality we observe.

12.2 Electromagnetic Fields in Matter

Electromagnetic fields are fundamental to the behavior of matter, shaping its interactions, transformations, and the intricate dance of forces that govern the physical world. Within the Unified Energy Dynamics (UED) framework, these fields emerge as active energy manifestations that connect atomic structures to the universal energy medium, enabling interactions at every scale—from chemical bonds to nuclear reactions.

Atomic Structures and Universal Fields

Atoms are not isolated entities; they interact continuously with the electromagnetic fields that permeate the universe. These interactions define the properties and behaviors of matter:

Electron Clouds and Field Dynamics: Electrons in atoms exist within quantized energy levels, their movements creating localized electric fields. These fields interact with the universal energy medium, enabling phenomena like photon absorption, emission, and wave-particle duality.

Proton Influence: The positive charge of protons within the nucleus establishes a counterbalance to the electron cloud, generating electromagnetic forces that stabilize the atom while also interacting with external fields.

Electromagnetic Forces: The Architects of Interaction

Electromagnetic forces, driven by the interaction of charges, are central to both chemical and nuclear processes:

Chemical Reactions: The bonds that form and break between atoms are governed by electromagnetic interactions between electron clouds. When atoms share or transfer electrons, the energy transitions involved are facilitated by the electromagnetic fields they generate.

Nuclear Reactions: Within the nucleus, the interplay between protons and neutrons is influenced by electromagnetic forces. Although the strong nuclear force holds the nucleus together, the repulsive electromagnetic force between protons determines the stability of the nucleus and its potential for energy release during fission or fusion.

Energy Dynamics in Electromagnetic Fields

Electromagnetic fields serve as conduits for energy transfer within and between atoms:

Energy Absorption and Emission: When atoms interact with external electromagnetic fields, their electrons absorb or release energy, resulting in transitions between energy states. These transitions drive phenomena like fluorescence, radiation, and even photosynthesis.

Field Interactions in Matter: The interaction of atomic fields with external electromagnetic waves underpins technologies like spectroscopy, enabling the detection and analysis of matter's properties.

Universal Relevance of Electromagnetic Fields

Electromagnetic fields are not confined to localized interactions; they link matter to the infinite energy pool:

Fields as Universal Extensions: Electric and magnetic fields originating from atomic structures align with the universal energy medium, ensuring coherence and continuity in energy dynamics.

Cosmic Implications: Electromagnetic fields generated by matter influence large-scale structures, from planetary magnetospheres to interstellar radiation, highlighting their universal significance.

The Unifying Role of Electromagnetic Forces

Electromagnetic fields are the unseen architects of matter's behavior, governing the balance between stability and dynamism. By mediating energy transitions and enabling interactions, these fields exemplify the interplay between active energy and the stable energy medium, revealing a universe interconnected at every level. Through their action, matter becomes a vibrant participant in the universal symphony of energy.

12.3 Energy Transitions in Nuclear Reactions

Nuclear reactions represent some of the most profound manifestations of energy transitions, where the concentrated active energy confined within atomic nuclei is released, transformed, and reabsorbed into the energy medium. Within the framework of Unified Energy Dynamics (UED), these reactions illustrate the interplay between stability and dynamism, providing insights into the nature of matter and the energy it embodies.

Active Energy Release in Fission, Fusion, and Radioactive Decay

The release of active energy in nuclear reactions highlights the immense potential contained within matter:

Fission: Splitting the Atom's Stability

In nuclear fission, a heavy nucleus splits into two or more smaller nuclei when struck by a neutron or other particle. This process releases a significant amount of energy as the nucleus transitions from a highly concentrated energy state to more stable configurations.

The energy released during fission arises from the mass defect, where a portion of the nucleus's mass is converted into energy, aligning with Einstein's equation $E=mc^2$.

Examples include the reactions in nuclear reactors and atomic bombs, where controlled and uncontrolled fission processes harness or unleash this energy.

Fusion: Creating Energy Through Unity

Fusion involves the merging of lighter nuclei, such as hydrogen isotopes, into a heavier nucleus, like helium. This process occurs in extreme conditions, such as the cores of stars, where immense pressure and temperature overcome the repulsive forces between nuclei.

Fusion reactions release energy as the resulting nucleus has less mass than the combined mass of the original nuclei, with the difference converted into active energy.

The energy produced powers stars and drives phenomena like solar radiation, forming the foundation of life and energy distribution in the cosmos.

Radioactive Decay: The Gradual Transformation

In radioactive decay, unstable nuclei spontaneously emit particles and energy to transition to more stable states. Processes like alpha decay, beta decay, and gamma emission release active energy while transforming the nucleus's composition.

The decay is governed by intrinsic instability within the nucleus, reflecting the principles of UED where stability drives transitions.

Mass Defect and Energy Dynamics

The mass defect—the difference between the mass of a nucleus and the sum of its constituent protons and neutrons—provides critical evidence of UED principles:

Conversion of Mass to Energy: The mass defect arises because some of the mass is converted into binding energy, which holds the nucleus together. This binding energy is a form of active energy, stored within the nucleus and released during fission or fusion.

UED interprets this as the localized concentration of energy transitioning back into the stable energy medium, completing the energy cycle.

Unified Energy Perspective: Mass is not an intrinsic property but a manifestation of active energy condensed within the nucleus. The mass defect illustrates how active energy integrates with stable energy to create physical matter.

During nuclear reactions, this concentrated energy is liberated, contributing to the dynamic equilibrium of the universal energy medium.

Implications for UED Principles

Nuclear reactions exemplify the dynamic interplay of stability and motion central to UED:

Energy Flow and Transformation: The transitions in fission, fusion, and decay reflect the cyclical nature of energy dynamics, where active energy is continuously transformed and reintegrated into the stable energy pool.

These processes underline the principle that energy is neither created nor destroyed but transitions between forms and states.

Cosmic Scale of Nuclear Energy: The energy released during nuclear reactions fuels stars, powers cosmic phenomena, and sustains the energy flow in the universe. It demonstrates the role of active energy in shaping the cosmos, from subatomic particles to galactic structures.

Nuclear reactions serve as a vivid illustration of UED principles, showcasing the profound connections between mass, energy, and stability. By examining these processes through the lens of UED, we gain a deeper understanding of the energy transitions that drive the universe, revealing the harmony underlying even the most explosive phenomena.

12.4 Quantum-Level Vibrations: Sustaining Active Energy

In the realm of quantum mechanics, matter ceases to be static and reveals a constant state of vibratory motion. These quantum-level vibrations are not incidental but represent the fundamental mechanism by which active energy remains connected to the stable energy pool. Through these oscillations, matter sustains its dynamic existence, ensuring its interactions within the universal framework of Unified Energy Dynamics (UED).

Vibratory Motion: The Universal Signature of Active Energy

Inherent Motion at the Quantum Scale: At the quantum level, particles such as electrons, protons, and even entire atomic nuclei exhibit vibratory and oscillatory motion. These vibrations are manifestations of active energy, continuously in flux yet grounded in the stable energy medium.

Vibrations are not restricted to particles in isolation; they extend to entire molecular structures, where vibrational modes define the physical and chemical properties of matter.

Energy Retention Through Oscillation: Vibratory motion ensures that active energy remains confined within matter while maintaining a connection to the stable energy pool.

This oscillatory behavior is not chaotic; it follows well-defined principles that regulate energy levels, transitions, and the interactions of particles and waves.

Quantum Vibrations and the Connection to the Energy Pool

Oscillations as a Bridge: Vibrations serve as the bridge between active energy within matter and the stable energy pool, allowing seamless energy flow and transformation.

For instance, when electrons oscillate between energy states, the energy emitted or absorbed is directly linked to the stable pool, showcasing the interconnectedness of quantum dynamics and the universal medium.

Energy Dissipation and Stability: Quantum vibrations also govern energy dissipation. When matter undergoes transformations, such as during chemical reactions or radiative processes, the vibratory energy facilitates the redistribution of active energy back to the stable pool.

This cyclical process underscores the principle of energy conservation in UED, where no energy is lost, only transitioned between states.

The Role of Vibratory Motion in Matter's Stability

Defining Matter's Structural Integrity: Vibratory motion at the quantum level dictates the structural stability of matter. The vibrational frequencies of atomic bonds, for example, determine molecular rigidity, flexibility, and reactivity.

These vibrations ensure that matter retains coherence while remaining dynamic, enabling transformations without compromising stability.

Quantum Vibrations as Dynamic Anchors: While matter appears stationary at macroscopic scales, quantum vibrations act as anchors that maintain its connection to the universal energy medium.

These motions are not arbitrary; they are governed by intrinsic energy states, quantum mechanical principles, and interactions with the surrounding energy field.

Implications for UED and Universal Dynamics

A Universal Mechanism: Vibratory motion exemplifies the universality of active energy dynamics. Whether in subatomic particles, complex molecules, or cosmic structures, oscillations sustain the equilibrium between stability and motion.

These vibrations are not isolated phenomena; they are intrinsic to the entire cosmos, uniting micro and macro realms through shared principles.

Energy Dynamics and Creation: Vibratory motion plays a critical role in the creation of physical matter. Disturbances in the stable energy pool give rise to active energy, which manifests as particles vibrating within defined quantum states.

This process not only sustains the existence of matter but also facilitates its evolution, ensuring adaptability and responsiveness within a dynamic universe.

Quantum Vibrations: A Symphony of Universal Harmony

The Interplay of Frequency and Stability: Every vibratory motion is characterized by a specific frequency, which acts as a signature of the particle or system's energy state. This frequency reflects the balance between active energy and the stabilizing influence of the universal energy medium.

Cosmic Resonance Through Vibrations: On a cosmic scale, vibrations contribute to the resonance observed in celestial bodies and fields. This resonance exemplifies the harmonious interplay of forces and energies, maintaining the coherence of the universe.

Quantum-level vibrations are not merely microscopic phenomena; they are the fundamental expressions of active energy, sustaining matter's existence and its relationship with the stable energy pool. By understanding these vibrations through the lens of UED, we gain insights into the intricate mechanisms that connect the smallest particles to the vast cosmic web, revealing a universe unified by the eternal dance of energy.

12.5 Fields without Matter: Evidence of the Energy Medium

The presence of fields in seemingly "empty" space has long puzzled physicists and philosophers alike. These fields, such as electromagnetic and gravitational fields, exist independently of tangible matter, pervading vast expanses of the universe. Through the lens of Unified Energy Dynamics (UED), these fields are not anomalies but clear evidence of the stable energy pool—the unseen medium that forms the foundation of existence.

The Origin of Fields in "Empty" Space

Disturbances in the Stable Energy Pool: Fields in space are not spontaneous occurrences but arise from localized disturbances in the stable energy medium. These disturbances, created by active energy or massive objects, ripple through the pool, generating fields that extend across space.

For instance, the gravitational field surrounding a massive object is the result of the object's interaction with the energy medium, creating a localized curvature or distortion that radiates outward.

Universal Presence of Fields: Even in regions devoid of matter, fields persist, suggesting that they are intrinsic to the energy medium itself. Electromagnetic waves, for example, propagate through space as oscillating electric and magnetic fields, without requiring a material substrate.

These fields act as the energy medium's manifestation, making its presence detectable even in the absence of physical matter.

Fields as the Unseen Fabric of Universal Coherence

Stabilizing the Cosmic Structure: Fields play a crucial role in maintaining universal coherence. They act as stabilizing agents that guide energy transitions, interactions, and motion at all scales—from the subatomic to the cosmic.

Gravitational fields ensure that celestial bodies follow predictable trajectories, while electromagnetic fields mediate interactions between charged particles, enabling the formation of matter and the transfer of energy.

Energy Flow Without Matter: Fields facilitate the flow of energy across the universe, even in the absence of physical carriers. This highlights their role as extensions of the energy medium, bridging the tangible and intangible aspects of reality.

For example, the electric and magnetic components of electromagnetic fields oscillate in harmony, allowing energy to propagate through space in the form of light or other electromagnetic waves.

Fields as Evidence of the Universal Energy Medium

A Reflection of the Energy Pool: Fields are not independent entities but direct reflections of the stable energy pool's dynamics. They embody the energy medium's capacity to sustain and transmit forces, linking all phenomena into a unified whole.

The existence of fields in empty space challenges traditional notions of a vacuum, suggesting instead that space is an active participant in the cosmic dance of energy and matter.

Connecting the Micro and Macro Realms: Fields serve as the invisible threads connecting the smallest particles to the largest celestial bodies. They bridge quantum phenomena, like electron transitions, with macroscopic effects, such as gravitational waves propagating across galaxies.

This interconnectedness aligns with UED's principle that all phenomena are unified by the dynamics of energy within the stable medium.

Fields Without Matter: A New Perspective

A Medium for Universal Coherence: Fields provide the scaffolding for universal interactions, ensuring that energy dynamics operate harmoniously. They reveal the energy medium's omnipresence and its role in shaping the structure and evolution of the cosmos.

Whether in the form of gravitational curvature, electromagnetic oscillations, or other manifestations, fields underscore the stability and dynamism inherent in the energy pool.

Implications for Understanding Reality: By recognizing fields as disturbances in the stable energy medium, UED offers a paradigm shift in understanding their origin and function. Fields are not secondary effects but primary components of the universe's fabric, integral to its coherence and balance.

Fields in "empty" space are profound indicators of the stable energy medium's presence. They are the energy pool's dynamic expressions, facilitating interactions and sustaining the universe's intricate web of relationships. Through UED, we understand these fields not as anomalies but as the unseen fabric of reality, uniting the cosmos in an elegant and harmonious system.

12.6 Unifying Atomic and Cosmic Energy Dynamics

The universe, vast in scale and complexity, is governed by principles that seamlessly link the smallest particles to the largest celestial structures. Through the lens of Unified Energy Dynamics (UED), the apparent divide between atomic phenomena and cosmic events dissolves, revealing a cohesive framework where the same energy principles govern both realms. This interconnectedness underscores the elegance of the universe's design, where the micro and macro harmonize in a perpetual dance of energy transitions.

Linking Atomic Phenomena to Universal Principles of UED

Energy Dynamics Across Scales: At the atomic level, particles like protons, neutrons, and electrons embody active energy, constantly interacting with the stable energy medium to maintain their structure and behavior. These interactions result in phenomena such as chemical bonding, nuclear reactions, and the emission of photons.

On a cosmic scale, the same energy principles drive the motion of celestial bodies, the formation of galaxies, and the propagation of gravitational waves. UED unites these processes under the umbrella of energy interactions, emphasizing that scale does not alter the fundamental dynamics.

Vibratory Motion and Stability: Vibratory motion, a hallmark of active energy at the atomic level, ensures the connection between matter and the stable energy pool. This same principle of stability amidst motion governs the orbiting of planets and the rotation of galaxies, creating a parallel between the atomic and cosmic realms.

UED highlights this vibratory motion as a universal signature of energy, bridging the smallest particles to the grandest cosmic structures.

How the Behavior of Particles Mirrors Cosmic Harmony

Interconnected Forces: The electromagnetic force, crucial for atomic interactions, finds its counterpart in the large-scale behavior of electromagnetic waves that traverse interstellar space. Similarly, gravitational forces, subtle at atomic scales, shape the architecture of the cosmos, binding galaxies into harmonious structures.

These forces are not isolated phenomena but interconnected expressions of the energy medium's dynamics, weaving a consistent thread through the universe's fabric.

Cycles of Creation and Dissolution: At the quantum level, particles engage in energy transitions, emitting or absorbing photons during chemical or nuclear reactions. These processes mirror the cosmic cycles of star formation, supernovae, and black hole activity, where energy is continuously transformed and reintegrated into the stable pool.

UED reveals that the behavior of particles is a microcosm of cosmic evolution, with both realms driven by the same principles of energy conservation, transformation, and stability.

Patterns and Predictability: The predictable behavior of atomic particles, guided by quantum mechanics, reflects the orderly motion of celestial bodies governed by gravitational and electromagnetic forces. These patterns emerge from the intrinsic harmony of the energy medium, ensuring coherence across scales.

The cyclical nature of energy transitions at both levels underscores the universe's inherent order, where chaos is balanced by stability, and randomness is tempered by universal laws.

The Unified Perspective of UED

From Quarks to Galaxies: By connecting the behavior of atomic particles to the dynamics of cosmic structures, UED offers a unified perspective that transcends traditional divisions. The principles that govern quarks, protons, and electrons are the same that shape stars, galaxies, and the vast expanse of the cosmos.

This unity reflects the energy medium's role as the foundation of existence, linking all phenomena into a single, coherent framework.

Bridging the Micro and Macro: UED emphasizes that the universe is not fragmented into separate domains but is a continuous system where the micro and macro are intimately connected. The same energy transitions that sustain atomic particles drive the motion of galaxies, illustrating the universality of energy dynamics.

The unification of atomic and cosmic energy dynamics through UED highlights the profound interconnectedness of the universe. From the tiniest particles to the largest celestial bodies, all phenomena emerge from the same principles of energy interaction and transformation. This perspective not only deepens our understanding of the cosmos but also affirms the harmony underlying its infinite complexity, demonstrating that the universe, in all its vastness, operates as a unified whole.

Chapter 13: Momentum and Matter Waves in Energy Dynamics

In the grand tapestry of the universe, motion is the essence of existence, and momentum is its silent orchestrator. From the ceaseless dance of particles in the quantum realm to the majestic revolutions of celestial bodies, momentum serves as the universal language of motion. It is not merely a physical quantity; it is an expression of active energy in motion, a fundamental principle that links the microcosm of atomic behavior to the macrocosm of cosmic dynamics.

In this chapter, we explore momentum as a cornerstone of energy dynamics, delving into its dual manifestations in particles and waves. Through the lens of Unified Energy Dynamics (UED), we uncover how momentum and matter waves are not isolated phenomena but integral components of a harmonious and interconnected universe. By understanding their roles and interactions, we gain deeper insights into the unified principles that govern all motion, transformation, and equilibrium.

13.1 Momentum: The Engine of Motion

Momentum is one of the most fundamental concepts in physics, representing the very essence of motion. It is often described as the product of an object's mass and velocity in classical mechanics, symbolizing the persistent nature of motion. Yet, in the quantum realm, momentum takes on a more nuanced form, manifesting as a probabilistic property tied to a particle's wave-like behavior. Regardless of the scale, momentum is the engine that drives motion, bridging the tangible world of observable phenomena with the intangible dynamics of energy.

In the framework of Unified Energy Dynamics (UED), momentum is not just a property of mass in motion but an expression of active energy engaging with the stable energy pool. It is the direct result of energy transitions, where stable energy transforms into dynamic forms, propelling particles, waves, and even celestial bodies. This perspective broadens our understanding of momentum, positioning it as a universal principle that governs all scales of existence, from the motion of subatomic particles to the grand revolutions of galaxies.

By redefining momentum within UED, we see it as more than a mechanical quantity—it is the tangible evidence of energy's perpetual flow, ensuring coherence and continuity in the universe's symphony of motion.

Matter Waves: Particles in Motion

MATTER WAVES ARE A profound illustration of the universe's inherent duality, encapsulating both the particle-like and wave-like nature of existence. This wave-particle duality, famously demonstrated by quantum experiments, reveals that every particle in motion exhibits wave-like properties, a phenomenon that challenges the classical view of matter as merely solid and localized.

The concept of matter waves was first introduced by Louis de Broglie, who proposed that particles possess a wavelength inversely proportional to their momentum, now known as the de Broglie wavelength. This revolutionary idea bridged the gap between classical mechanics and quantum theory, highlighting that the behavior of matter is governed by universal principles of energy dynamics. The de Broglie wavelength, though imperceptibly small for macroscopic objects, becomes significant for microscopic particles, providing experimental evidence for the interconnectedness of energy and matter.

In the context of Unified Energy Dynamics (UED), matter waves are not just mathematical abstractions; they are the signatures of active energy in motion, maintaining a connection to the stable energy pool. The wave-like properties of particles reflect their continuous interaction with the energy medium, a dynamic interplay that ensures coherence across scales. Matter waves embody the essence of UED: they demonstrate that all motion, whether of particles or celestial bodies, is a manifestation of energy transitions within the infinite medium.

By understanding matter waves through this lens, we gain deeper insight into how the universe maintains its harmony, connecting the micro and macro realms in a unified framework of motion and transformation.

13.2 Momentum Conservation in UED

The principle of momentum conservation is a cornerstone of both classical and quantum physics, signifying that the total momentum of a system remains constant in the absence of external forces. Within the framework of Unified Energy Dynamics (UED), this principle is elevated to a universal law that operates across all scales, seamlessly bridging the micro and macro realms of the cosmos.

Momentum in the Stable Energy Pool

In UED, momentum is viewed as a direct expression of active energy in motion, intimately tied to the stable energy pool. This universal medium acts as the backdrop where energy transitions and interactions occur, ensuring that momentum is neither created nor destroyed but continuously redistributed. The stable energy pool absorbs disturbances and mediates interactions, maintaining equilibrium even as individual components undergo dynamic transformations.

For instance:

At the atomic level, particles like electrons exhibit momentum as they orbit the nucleus, maintaining a delicate balance between attractive and repulsive forces.

At the cosmic scale, celestial bodies conserve momentum as they revolve around massive objects, their motion sustained by the dynamic equilibrium of the energy medium.

Bridging Micro and Macro Systems

Momentum conservation within UED highlights the interconnectedness of phenomena across scales. The same principles that govern the trajectory of an electron within an atom also dictate the movement of planets and galaxies. This universality arises from the continuous interaction of active energy with the stable pool, ensuring that momentum transitions remain consistent regardless of scale.

Key examples include:

Subatomic Interactions: In particle collisions, momentum is conserved as energy transitions between forms, demonstrating the stability of the underlying energy framework.

Gravitational Dynamics: The orbits of celestial bodies reflect the conservation of angular momentum, with the energy medium acting as the unseen stabilizer of motion.

A Universal Bridge

Momentum conservation serves as a bridge between the tangible and intangible, the measurable and the metaphysical. By anchoring all motion within the stable energy pool, UED provides a unified understanding of how momentum sustains the coherence of the universe. Every interaction, from the simplest particle collision to the intricate dance of galaxies, reaffirms the principle of conservation, revealing a cosmos bound by dynamic harmony.

Through UED, momentum is not just a property of moving objects but a fundamental characteristic of existence itself, ensuring that the universal interplay of forces and energies remains balanced and eternal.

13.3 Interplay of Momentum and Energy Transitions

Momentum and energy are two inseparable aspects of motion, intricately linked through the transformations and interactions that define the universe. Within Unified Energy Dynamics (UED), momentum is more than a measure of motion—it is a driving force that governs energy transitions, shaping phenomena across both atomic and cosmic scales.

Momentum: A Catalyst for Energy Transformations

Momentum serves as the dynamic engine that facilitates energy transitions, bridging stability and motion. Every interaction within the energy medium, from particle collisions to gravitational waves, involves a transfer of momentum that propels transformation and evolution. This interplay ensures the continual flow of active energy while preserving equilibrium within the stable energy pool.

Key insights include:

At the Quantum Scale: The momentum of subatomic particles, such as electrons, is instrumental in phenomena like quantum tunneling and orbital transitions. These processes involve the transfer of energy and momentum, enabling particles to shift states or traverse barriers, revealing the profound interconnectedness of motion and energy.

At the Cosmic Scale: Momentum sustains the orbits of celestial bodies and the propagation of gravitational waves, ensuring the coherence of large-scale structures. For example, the momentum transfer in binary star systems generates ripples in the energy pool, illustrating the universality of this principle.

Momentum Transfer Across Scales

The transfer of momentum is not confined to a single scale but occurs across the spectrum of existence, unifying micro and macro realms under a common framework.

Atomic Interactions: In chemical reactions, momentum transfer facilitates the rearrangement of bonds and the release or absorption of energy. Similarly, in nuclear reactions, such as fission and fusion, the redistribution of momentum governs the behavior of particles and the immense energy release.

Cosmic Dynamics: Planetary systems conserve angular momentum as they interact with gravitational fields, ensuring orbital stability. The collision of celestial objects, such as asteroids or galaxies, results in momentum transfer that reshapes cosmic landscapes, demonstrating the resilience of universal harmony.

Momentum as a Universal Driver

Momentum acts as the universal driver of energy dynamics, ensuring that every transformation is guided by the principles of conservation and coherence. Whether it is the spin of an electron or the rotation of a galaxy, momentum enables the flow of active energy through the stable pool, sustaining the perpetual dance of creation and dissolution.

In UED, the interplay of momentum and energy transitions is not merely a mechanical process but a manifestation of the universe's inherent harmony. By examining this relationship, we uncover the mechanisms that govern motion at all levels, revealing a cosmos where the smallest particle and the largest structure are unified by the eternal principles of energy dynamics.

13.4 Wave Mechanics and Energy Medium

Wave mechanics, a cornerstone of quantum physics, gains profound depth through the lens of Unified Energy Dynamics (UED). Matter waves, a manifestation of the wave-particle duality, represent the vibratory motion of active energy within the stable energy medium. These interactions between matter waves and the infinite energy medium illuminate the unified principles governing atomic and cosmic behaviors alike.

Interaction of Matter Waves with the Energy Medium

Matter waves, as described by the De Broglie hypothesis, are the oscillatory expressions of particles in motion. While traditionally considered isolated phenomena, UED reveals their intimate connection with the energy medium:

Vibrational Anchoring: Matter waves are not isolated oscillations but are sustained by the stable energy medium. The energy pool acts as both the anchor and the backdrop, enabling particles to maintain coherence and wave-like behavior while navigating their environments.

Propagation Through Stability: Unlike mechanical waves, which rely on particle interactions in a material medium, matter waves propagate by interacting with the inherent stability of the energy pool. This interaction sustains their wave-like nature across different scales, from electrons in atoms to particles traveling through interstellar space.

Atomic-Level Implications

The interplay of matter waves with the energy medium governs a range of atomic behaviors:

Quantum Stability: The stability of atomic orbitals is a direct consequence of the balance between matter wave oscillations and the energy medium's stabilizing influence. This balance ensures that electrons remain bound yet dynamic, enabling chemical bonds and reactions.

Energy Transitions: When electrons transition between energy levels, their wave properties shift, releasing or absorbing photons. This phenomenon, a direct result of matter wave interactions, demonstrates the seamless integration of wave mechanics with energy dynamics.

Cosmic Implications

The principles governing matter waves extend beyond the atomic scale, influencing larger cosmic structures:

Interstellar Motion: Particles traveling through the vastness of space retain their wave-like properties due to the energy medium. These matter waves interact with the energy pool, guiding their trajectories and ensuring coherence over astronomical distances.

Gravitational Wave Parallels: Just as gravitational waves ripple through spacetime, matter waves ripple through the energy medium, creating a bridge between quantum behaviors and cosmic dynamics.

The Unified Framework of Wave Mechanics

Wave mechanics, when viewed through the UED framework, offers a unifying thread between the micro and macro realms:

Atomic Harmony: The energy medium ensures the stability of atomic systems, facilitating the emergence of matter's diverse properties and behaviors.

Cosmic Connectivity: The same principles that govern matter waves at the atomic level extend to celestial phenomena, highlighting the universal applicability of wave mechanics.

By redefining wave mechanics within the context of the energy medium, UED provides a cohesive understanding of matter waves as integral components of the universe's dynamic equilibrium. This perspective not only enriches our understanding of quantum phenomena but also reinforces the interconnectedness of all existence.

13.5 The Unified Perspective of Momentum

Momentum, long studied within the frameworks of classical mechanics and quantum physics, emerges as a universal property that bridges energy and matter in Unified Energy Dynamics (UED). Beyond its conventional definitions, momentum is redefined in UED as the active expression of energy in motion, seamlessly integrating dynamics across micro and macro realms.

Momentum in Classical, Quantum, and Universal Contexts

In classical mechanics, momentum is a straightforward product of mass and velocity, describing the motion of objects within localized systems. Quantum mechanics, however, reveals a more nuanced view: momentum is not only tied to particle motion but also intrinsic to wave-like behaviors, as encapsulated in de Broglie's wavelength.

Classical Momentum: Governs the motion of macroscopic objects, ensuring the conservation of motion in collisions and interactions. For example, the momentum of a moving car or a pendulum reflects the laws of conservation within a terrestrial, localized framework.

Quantum Momentum: Unveils the dual nature of particles, such as electrons, which exhibit both wave-like and particle-like properties. The Heisenberg Uncertainty Principle, linking momentum and position, emphasizes the probabilistic nature of quantum states.

UED's Universal Momentum: Extends the concept beyond the constraints of mass and velocity to encompass the flow of active energy within the infinite stable pool. In UED, momentum becomes the connective thread that binds localized phenomena to the broader universal framework, where stability and dynamism coexist.

Momentum: Binding Energy and Matter

Momentum serves as the dynamic bridge between energy and matter, ensuring coherence in both their transformations and interactions.

In Atomic Systems: Momentum drives quantum phenomena, such as electron transitions between energy levels. The exchange of momentum in these transitions facilitates the emission and absorption of photons, underscoring the active energy exchanges central to atomic behavior.

In Cosmic Systems: Momentum governs the motion of celestial bodies, from the spin of planets to the rotation of galaxies. Gravitational waves, as ripples in the energy medium, propagate momentum across vast cosmic distances, maintaining equilibrium and structure.

The Universal Property of Momentum

Within UED, momentum transcends its localized definitions, emerging as a universal property that reflects the interplay of energy and matter. It is not confined to individual particles or objects but is inherent to the fabric of the energy medium itself.

Key insights include:

Momentum in the Energy Medium: The stable energy pool acts as both a reservoir and conduit for momentum, ensuring its conservation across scales. This perspective unifies the behavior of particles, waves, and celestial bodies under a single, cohesive framework.

Momentum's Role in Universal Harmony: Momentum facilitates the balance between stability and motion, ensuring that transformations occur in a manner that sustains cosmic harmony. Whether in the vibration of an atom or the orbit of a planet, momentum maintains the coherence of the universe's dynamic processes.

A Unified Perspective

Thus, Momentum, as redefined by UED, is more than a measure of motion; it is a universal principle that binds energy and matter, micro and macro, stability and dynamism. By uniting classical mechanics, quantum physics, and universal energy dynamics, momentum becomes a cornerstone of the interconnected cosmos. This unified perspective not only deepens our understanding of physical phenomena but also illuminates the profound harmony that underpins the universe.

Conclusion: Momentum as Nature's Dynamic Pulse

Momentum, often perceived as a straightforward quantity in classical and quantum physics, emerges in Unified Energy Dynamics (UED) as a profound expression of the universe's dynamic pulse. It encapsulates the active energy in motion, weaving through the cosmic and quantum realms as a binding force that sustains harmony across scales. From the smallest particles to the largest celestial bodies, momentum is the silent orchestrator of motion, interaction, and transformation.

Within the framework of UED, momentum transcends its traditional boundaries, unifying the behaviors of energy and matter. It is not merely a product of mass and velocity but a manifestation of energy transitions within the infinite energy medium. This perspective deepens our understanding of momentum's conservation, its role in energy transfer, and its influence on the structural and dynamic integrity of the universe.

Momentum is the engine that drives the dance of creation, sustaining the interplay of stability and dynamism essential for cosmic harmony. Through this lens, we discover momentum as more than a measurable quantity—it is a universal principle reflecting the ceaseless rhythm of energy and matter in motion. As a cornerstone of UED, momentum connects the tangible and intangible, reaffirming the unity of existence within the boundless energy medium.

Summary of Part III

Micro and Macro Energy Dynamics

In Part III, we delve deep into the intricate relationship between energy dynamics at micro and macro scales, illustrating how the principles of Unified Energy Dynamics (UED) bridge the smallest quantum interactions to the vast structures of the cosmos. This section establishes a cohesive understanding of energy as the unifying thread across all phenomena.

The Photon and Energy Medium: This chapter explores photons as carriers of universal energy, embodying the dynamic balance between wave and particle natures. The interaction of photons with the stable energy pool reveals the intricate mechanism of energy propagation without the need for a traditional medium. Their lifecycle—from emission to reintegration into the energy medium—offers insights into the eternal cycles of energy in motion.

The Mechanics of Waves in the Energy Medium: Here, we contrast electromagnetic and mechanical waves, highlighting the unique ability of electromagnetic waves to propagate through the infinite energy medium. By comparing wave behavior across scales and examining their propagation in the cosmic energy pool, this chapter unveils the role of waves as universal energy connectors, transcending material constraints.

Cosmic Harmony: Unified Energy in Action: This chapter synthesizes the micro and macro perspectives, showcasing how energy interactions govern phenomena from atomic particles to galaxies. Gravitational waves are revisited as ripples within the energy medium, while the interplay between stability and motion across cosmic structures emphasizes the dynamic balance that sustains the universe's harmony.

Through Part III, UED reveals a unified framework for understanding the interconnectedness of all energy phenomena. It provides a lens to view the universe as an integrated whole, driven by the harmonious interplay between stable and active energy. From photons and waves to galaxies and gravitational forces, this section emphasizes the profound unity underlying the infinite complexity of existence.

Brainstorming Session

REFLECTING ON THE FOUNDATIONS of UED

At the conclusion of these three foundational parts, it's vital to pause and reflect on the revolutionary ideas presented in Unified Energy Dynamics (UED). This brainstorming session invites readers to challenge, explore, and expand the principles discussed so far, bridging science, philosophy, and intuition.

Questions to Ponder

The Nature of Energy:

If energy is the essence of existence, can we identify phenomena that don't fit into the stable-active energy framework?

Could the stable energy pool contain unknown states of energy yet to be observed?

Time and Space:

If time is merely a localized perception, how do we reconcile this with measurable cosmic events, like supernovae or the expansion of the universe?

Could space, as the infinite energy medium, have properties that challenge its definition as "empty" or "vacuum"?

The Role of Photons and Waves:

Are there other carriers of energy in the universe besides photons?

Could we one day detect the vibratory nature of energy at quantum and cosmic levels directly?

Gravity and Cosmic Harmony:

If gravity is not a force but a manifestation of mass-energy interactions, how might this redefine our approach to understanding black holes or galaxy formation?

Could gravitational waves be used to map the energy medium itself?

Atomic and Quantum Dynamics:

How does the behavior of particles like electrons and protons reflect larger cosmic principles?

Are fields in "empty space" evidence of a more complex structure of the energy pool than we currently comprehend?

Expanding UED: The Path Forward

Bridging Gaps:

How can UED be integrated with existing scientific theories, like relativity and quantum mechanics, to create a more holistic framework?

Are there mathematical models or experimental methods that could validate UED's principles?

Philosophical Implications:

If UED redefines consciousness as energy in transit, what does this mean for our understanding of self-awareness, intelligence, and life?

Could UED provide insights into metaphysical questions about the origin and purpose of existence?

Practical Applications:

How might understanding energy dynamics at this level transform energy production, communication systems, or even space exploration?

Could new technologies arise from our ability to harness stable-active energy transitions directly?

Call to Action for Readers

Challenge: Examine the principles of UED in the context of real-world phenomena and existing scientific frameworks.

Explore: Investigate unanswered questions, propose experiments, or develop models that could extend or refine UED.

Reflect: Consider how UED aligns with or challenges your understanding of reality, and what new perspectives it opens up.

This brainstorming session aims not just to solidify the concepts presented but to inspire curiosity and dialogue, pushing the boundaries of our understanding of the universe. It's an invitation to participate in shaping the ongoing journey of UED, turning questions into discoveries and ideas into possibilities.

Part IV: Consciousness and Universal Connections

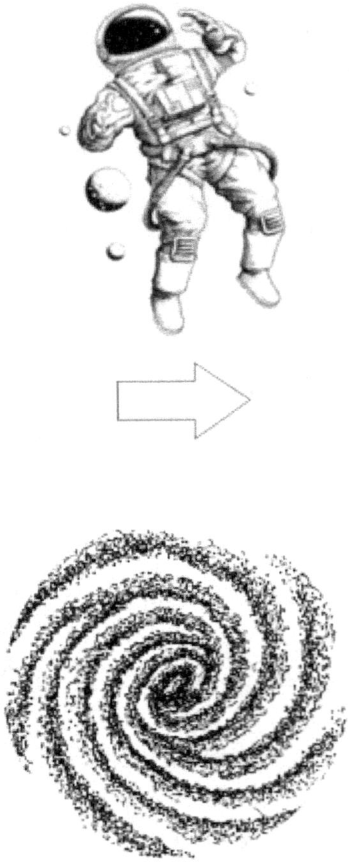

Chapter 14: Consciousness: Energy in Transit

In the journey through Unified Energy Dynamics (UED), we have explored the intricate interplay of forces, waves, and matter within the boundless energy medium. Yet, one profound phenomenon remains at the heart of existence—consciousness. Far beyond mere biological function, consciousness emerges as the dynamic bridge between stable and active energy, connecting the tangible and intangible realms of reality. From the deliberate thought processes of humans to the quantum-level awareness in matter, consciousness reflects the universe's innate ability to perceive, interact, and evolve.

This chapter delves into the essence of consciousness as energy in transit—a phenomenon that transcends individual perception and integrates with the universal energy pool. By examining the spectrum of awareness, from human intellect to the subtle behaviors of quantum particles, we uncover how energy dynamics shape the very fabric of perception and existence.

14.1 Introduction: Consciousness as the Pulse of Existence

Consciousness stands as one of the most enigmatic and essential phenomena in the universe, a dynamic force that resonates throughout the principles of Unified Energy Dynamics (UED). More than a mere byproduct of biological processes, consciousness emerges as the pulse of existence, a phenomenon intrinsically tied to the interaction between stable and active energy. It transcends individual perception, weaving through the fabric of reality as a universal thread that connects the micro and macro realms.

In UED, consciousness is understood as energy in transit—a phenomenon that bridges stability and motion, shaping the observable and intangible dimensions of the cosmos. It is not confined to the intellectual awareness of humans but extends to the subtle responsiveness of quantum particles and the organized coherence of celestial systems. Consciousness reflects the universe's innate ability to observe, adapt, and evolve, offering insights into the profound interplay between energy states.

This chapter explores the role of consciousness within UED, highlighting its ability to bridge energy states and influence the construction of reality. By examining its manifestations across scales, we uncover the mechanisms through which consciousness emerges and interacts with the universal energy medium. Through this lens, consciousness is not only a mirror of existence but also a force that actively shapes the evolution of the cosmos.

14.2 The Bridge between Stability and Dynamism

Consciousness is the bridge that connects the two fundamental states of energy—stability and dynamism—within the framework of Unified Energy Dynamics (UED). It operates as the intermediary force that allows the silent potential of stable energy to interact with the transformative motion of active energy. In doing so, consciousness enables the universe to observe and influence its own evolution, creating awareness at every scale.

How Consciousness Links Stable and Active Energy

Stable energy represents the reservoir of latent potential, the unchanging foundation upon which the universe rests. Active energy, by contrast, drives motion, transformation, and observable phenomena. Consciousness emerges at the nexus of these states, serving as a dynamic transit point where stability meets motion.

Dynamic Perception and Action: In living systems, consciousness arises from the interaction of active energy within neural networks and the stability provided by the universal energy medium. This interaction forms the foundation of perception, thought, and deliberate action.

Universal Awareness: Even at the quantum level, consciousness manifests as the universe's ability to respond to disturbances in the stable energy medium. The superposition of quantum states and wave-particle duality reflect this subtle awareness, demonstrating how active energy interacts with its stable foundation.

The Role of Energy Transitions in Creating Awareness

Energy transitions are at the heart of consciousness, enabling the flow of awareness across scales. These transitions occur when stable energy interacts with disturbances, giving rise to active forms such as light, motion, and thought. This dynamic exchange creates the conditions for perception and interaction:

Quantum-Level Awareness: At the smallest scales, energy transitions drive phenomena such as quantum entanglement and superposition, where particles exhibit a form of awareness by responding to changes in their environment.

Biological Consciousness: In organisms, energy transitions within cells and neural systems create the complex networks needed for higher-order consciousness. Electrical signals, powered by energy shifts, form the basis of thought and memory, linking stable and active energy to human awareness.

Cosmic Awareness: On a macro scale, the flow of energy within stars, galaxies, and planetary systems reflects the universe's larger consciousness. The gravitational and electromagnetic forces guiding celestial bodies are expressions of how stability and dynamism interact to create cosmic awareness.

Consciousness as a Universal Principle

By linking stable and active energy, consciousness unifies the tangible and intangible, the physical and metaphysical. It is the mechanism through which the universe perceives itself and evolves. Whether in the quantum dance of particles, the neural networks of a human brain, or the gravitational symphony of galaxies, consciousness acts as the bridge that transforms energy into awareness, shaping the ever-changing reality of existence.

14.3 Human Consciousness: A Unique Manifestation

Human consciousness stands as one of the most intricate and profound expressions of energy dynamics within the universe. It is the culmination of countless interactions between stable and active energy, manifesting as thought, perception, and self-awareness. In the framework of Unified Energy Dynamics (UED), human consciousness represents a unique intersection where the soul, as dynamic awareness, and the mind, as its functional counterpart, collaborate to perceive, analyze, and influence reality.

The Mind as the Functional Manifestation of the Soul

The soul, in the context of UED, is the core of human dynamism—an active energy state that embodies awareness and potential. It is the essence of existence, drawing from the stable energy medium to create a vibrant and adaptable presence. The mind, by contrast, is the operational interface of this dynamic awareness, a structured network through which the soul engages with the physical world.

Dynamic Awareness: The soul acts as a reservoir of dynamic awareness, connecting humans to the universal energy pool. It provides the latent potential for creativity, intuition, and emotional depth, acting as the source of inspiration and inner purpose.

The Mind's Role: The mind translates the abstract awareness of the soul into functional outputs. Through neural networks and cognitive processes, it enables humans to process sensory information, solve problems, and make decisions. This collaboration between soul and mind creates a duality where abstract potential becomes concrete action.

Intellect and Perception as Outcomes of Energy Dynamics

Human intellect and perception are direct outcomes of energy dynamics, shaped by the intricate interplay of stability and motion within the brain and body. These attributes emerge as expressions of energy transitions, where stable energy supports coherence and active energy drives interaction.

Neural Networks and Energy Flow: The brain functions as a dynamic energy system, with electrical impulses and chemical signals acting as conduits for active energy. These interactions form the basis of memory, learning, and decision-making, linking physical structures with the intangible awareness of the soul.

Perception and Sensory Integration: Perception arises from the integration of energy inputs from the external environment. Photons, sound waves, and tactile stimuli are transformed into neural signals, allowing the mind to construct a coherent understanding of reality. This process exemplifies how stable and active energy collaborate to create conscious awareness.

The Intellect's Creative Power: Intellect is the culmination of energy transitions within the human mind, enabling abstract reasoning, critical thinking, and the capacity to imagine. By drawing from the stable energy pool, the intellect generates insights and innovations that transcend the immediate and observable.

Human Consciousness as a Nexus of Energy Dynamics

Human consciousness is a testament to the profound interconnectedness of energy states. It reflects the universe's ability to perceive and influence itself, condensed into a single, unique manifestation. The collaboration between the soul and mind, powered by energy transitions, creates a system that is both adaptive and introspective, capable of shaping not only individual lives but also the broader trajectory of existence.

Through the lens of UED, human consciousness is not an isolated phenomenon but a microcosm of the universal dance of energy. It embodies the principles of stability and dynamism, offering a glimpse into the infinite potential of the cosmos as it evolves through awareness and interaction.

14.4 Quantum Awareness: Subtle Forms of Consciousness

Consciousness is not confined to biological systems or the intellectual processes of humans; it is a universal phenomenon manifesting across all scales of existence. At the quantum level, matter exhibits a form of awareness that reflects the universe's intrinsic ability to perceive and respond. Within the framework of Unified Energy Dynamics (UED), this subtle awareness is a natural outcome of the interaction between particles and the infinite energy medium.

How Matter Exhibits Awareness at a Quantum Level

At its core, quantum awareness refers to the capacity of particles to "sense" and respond to changes in their environment. While not equivalent to human cognition, this behavior reflects an underlying awareness that connects even the smallest units of matter to the energy medium.

Wave-Particle Duality: Particles such as electrons and photons exhibit dual behaviors, oscillating between wave-like and particle-like states depending on how they are observed. This duality reflects an awareness of the conditions of measurement, suggesting that quantum entities are influenced by their interactions with the energy medium.

Quantum Superposition: In superposition, particles exist in multiple potential states simultaneously until observed or interacted with. This ability to hold multiple possibilities showcases a subtle form of awareness, where particles remain connected to the energy medium's potential states.

Quantum Entanglement: Entangled particles maintain a connection that allows them to instantaneously influence each other, regardless of distance. This phenomenon illustrates a deep level of interconnectivity and responsiveness, as entangled particles seem to "communicate" changes in their state, transcending spatial boundaries.

The Interaction of Particles with the Energy Medium

The behavior of particles at the quantum level is profoundly influenced by their interaction with the stable energy medium. This interaction provides the framework within which quantum phenomena occur, enabling particles to exhibit behaviors that mirror awareness.

Energy as the Context for Quantum Behavior: The stable energy pool acts as the medium through which quantum entities operate. It supports the transitions of particles between energy states, creating the conditions for superposition, entanglement, and wave-particle duality.

Localized Disturbances and Response: Quantum particles represent localized disturbances within the stable energy medium. Their behaviors—whether oscillating as waves or interacting as particles—reflect their ongoing relationship with this universal energy framework.

Quantum Measurement and the Collapse of Potential: When a quantum system is observed, its potential states collapse into a single reality. This interaction highlights the role of the energy medium as both a source of potential and a framework for the realization of possibilities.

Quantum Awareness as a Universal Principle

Quantum awareness is a subtle but profound manifestation of consciousness, illustrating the universe's inherent capacity for perception and response even at its smallest scales. Through their interactions with the energy medium, quantum particles embody the principles of stability and dynamism, bridging the intangible and observable aspects of existence.

By recognizing this awareness in matter, UED expands the concept of consciousness beyond biological and intellectual boundaries. Quantum awareness is not an isolated phenomenon but a reflection of the universal energy dynamics that connect all forms of existence, from subatomic particles to galaxies. It highlights the profound interconnectedness of the cosmos, where every particle and interaction contributes to the collective awareness that defines reality.

14.5 Energy Dynamics and the Nature of Perception

Perception is the mechanism through which consciousness interprets the world, transforming energy transitions into meaningful experiences. Within the framework of Unified Energy Dynamics (UED), perception emerges as a product of vibratory interactions between consciousness and the universal energy medium. It is the bridge that translates abstract energy dynamics into tangible reality, shaping the way beings experience existence.

How Consciousness Interprets Energy Transitions as Reality

At its core, consciousness interacts with the energy medium by interpreting energy transitions—shifts between stable and active states—as observable phenomena. This process transforms the abstract flow of energy into the concrete reality perceived by conscious beings.

Decoding Energy Inputs: Perception begins with the reception of energy signals, such as photons, sound waves, or tactile stimuli, which interact with sensory systems. These signals are interpreted as visual images, sounds, or textures, creating a coherent representation of the external world.

Translating Vibrations into Experience: The vibratory nature of energy transitions underpins all sensory experiences. For instance, light waves of varying frequencies are perceived as different colors, while sound waves create the sensations of pitch and tone. This translation reflects consciousness's ability to transform abstract vibrations into distinct, meaningful perceptions.

Synthesizing Reality: Consciousness integrates energy inputs across different sensory modalities to form a unified understanding of reality. This synthesis not only shapes how beings perceive the external world but also enables introspection and self-awareness, linking external stimuli to internal experiences.

Perception as a Product of Vibratory Interactions with the Energy Pool

The act of perceiving is deeply rooted in the interaction between consciousness and the vibratory nature of the universal energy pool. This interaction enables conscious beings to connect with the cosmos, interpreting its dynamics through their unique sensory and cognitive frameworks.

Vibratory Resonance: Perception arises from the resonance between the vibratory patterns of external stimuli and the internal vibratory states of consciousness. For example, the brain's neural oscillations align with external rhythms, such as light or sound waves, facilitating the interpretation of energy transitions.

Harmonic Connection: The stability of the energy pool provides the foundation for coherent vibratory interactions, ensuring that perceptions are not chaotic but harmonious. This harmonic connection enables beings to experience a stable and consistent reality, even as energy transitions occur dynamically.

The Role of Sensory Systems: Sensory systems act as intermediaries that translate the vibratory language of the energy pool into signals comprehensible to the mind. Photoreceptors in the eyes, for example, convert light waves into neural impulses, while mechanoreceptors in the skin respond to pressure and vibration.

Perception as a Dynamic and Universal Process

Perception is not a passive process but an active engagement with the energy medium. Consciousness continually interacts with the vibratory nature of energy transitions, shaping reality in ways that are both universal and unique to the perceiver.

Universality of Perception: Across all conscious beings, perception reflects the same fundamental principles of energy dynamics. Whether in humans, animals, or even at the quantum level, the interpretation of energy transitions shapes the experience of reality.

Individuality and Diversity: While grounded in universal energy principles, perception varies based on the unique structure and sensitivity of each organism's sensory and cognitive systems. This diversity enriches the tapestry of reality, allowing consciousness to experience existence in infinite ways.

Perception as the Mirror of Energy Dynamics

Through perception, consciousness transforms the abstract flow of energy into the concrete experiences that define reality. It is both a product and a reflection of the vibratory interplay between stability and motion within the energy pool. By interpreting energy transitions, perception not only connects conscious beings to the universe but also reveals the profound harmony underlying all existence.

Unified Energy Dynamics frames perception as a dynamic process, where consciousness actively participates in shaping reality. This perspective not only deepens our understanding of perception but also highlights its role as a universal phenomenon, bridging the tangible and intangible aspects of existence through the vibratory essence of energy.

14.6 The Universal Fabric of Consciousness

Consciousness is not limited to individual beings or isolated phenomena; it is a unifying principle woven into the very fabric of the cosmos. Within the framework of Unified Energy Dynamics (UED), consciousness emerges as a universal attribute, deeply interconnected with the dynamics of life, energy, and matter. It transcends the boundaries of physical forms, revealing the profound unity that underpins existence.

Consciousness as a Unifying Principle in the Cosmos

At its core, consciousness acts as the cohesive force that connects disparate elements of the universe into a harmonious whole. It is the bridge that links the tangible and intangible, enabling energy and matter to interact in meaningful ways.

A Continuum Across Scales: Consciousness is present across all scales of existence, from the quantum behaviors of particles to the intellectual processes of humans. At every level, it manifests as the capacity for perception, interaction, and response, unifying the micro and macro realms.

The Dynamic Thread of Awareness: Consciousness serves as the thread that binds energy dynamics to the observable world. By linking stable and active energy states, it enables the universe to perceive and influence its own evolution, creating a self-sustaining cycle of awareness and transformation.

Universal Awareness: Beyond individual beings, consciousness reflects the universe's innate ability to "know" and respond to itself. This universal awareness is evident in the ordered patterns of galaxies, the symmetry of natural laws, and the responsive behaviors of living organisms.

Insights Into the Interconnectedness of Life, Energy, and Matter

The universal fabric of consciousness highlights the deep interconnectedness of all existence. Life, energy, and matter are not separate entities but interdependent expressions of a single, unified reality.

Life as an Extension of Energy Dynamics: Life forms are dynamic systems that draw from the universal energy pool to sustain and evolve. Consciousness in living beings arises from the intricate interplay of energy within cells, neural networks, and sensory systems, demonstrating how life is a continuation of universal energy principles.

Energy as the Foundation of Interaction: Energy transitions drive all interactions, whether within atoms or between galaxies. Consciousness interprets and directs these transitions, transforming raw energy into the forces, processes, and patterns that shape reality.

Matter as a Manifestation of Awareness: Matter, as localized disturbances in the stable energy medium, exhibits a form of consciousness through its responsiveness to external forces and internal dynamics. This subtle awareness is evident in the behavior of particles, the cohesion of structures, and the adaptability of systems.

Consciousness as the Fabric That Binds

By recognizing consciousness as a unifying principle, UED reveals the profound harmony and coherence of the cosmos:

Interdependence of All Things: Consciousness underscores the interdependence of life, energy, and matter, highlighting their shared origins in the energy medium. This interconnectedness affirms that existence is not fragmented but seamlessly woven together.

A Dynamic Fabric: The universal fabric of consciousness is not static but dynamic, continuously evolving through the interactions of energy and matter. It reflects the perpetual motion of the cosmos, where stability and transformation coexist in balance.

A Framework for Understanding Existence: Consciousness provides a lens through which to understand the unity of the universe. It bridges the physical and metaphysical, offering insights into the purpose and interconnectedness of all forms of existence.

Consciousness as the Universal Thread

In short, Consciousness is the essence of unity in the cosmos, the thread that connects all aspects of existence into a single, cohesive fabric. Through its interplay with energy and matter, it reveals the profound interconnectedness of life and the universe. Unified Energy Dynamics frames consciousness not as an isolated phenomenon but as the pulse of the cosmos, a dynamic principle that sustains the harmony and evolution of reality itself.

14.7 Conclusion: Consciousness as Energy's Eternal Dance

Consciousness, as explored through the lens of Unified Energy Dynamics (UED), is far more than an abstract concept or a biological trait—it is the active expression of energy in its most profound and dynamic form. It embodies the universe's inherent ability to perceive, adapt, and evolve, reflecting the eternal dance between stability and motion, potential and realization.

Consciousness as the Active Expression of Energy

At its essence, consciousness is energy in motion, bridging the tangible and intangible, the finite and infinite. It serves as the mechanism through which the universe observes and influences itself, enabling existence to unfold with purpose and coherence.

Energy in Transit: Consciousness arises from the transitions of energy, transforming static potential into dynamic expression. It is the flow that connects the stable energy medium with the active forces shaping the cosmos.

The Self-Awareness of the Universe: Consciousness acts as the universe's self-reflective mirror, a phenomenon that allows energy and matter to engage in purposeful interactions. From quantum awareness in particles to the introspection of human minds, consciousness represents the universe's capacity for self-awareness.

Shaping the Universe's Dynamic Equilibrium

Consciousness plays a central role in maintaining and evolving the dynamic equilibrium of the universe. It harmonizes the interplay of forces, guiding the transitions that sustain the cosmos in balance.

Stability Meets Dynamism: By bridging stable and active energy, consciousness ensures the seamless integration of potential and transformation. It allows the universe to remain grounded while constantly evolving.

A Creative Force: Consciousness is the creative engine of the cosmos, driving innovation and adaptation across all scales of existence. It shapes the patterns of galaxies, the behavior of particles, and the evolution of life itself.

The Equilibrium of Awareness: In its essence, consciousness seeks balance—a dynamic state where stability and motion coexist. This equilibrium sustains the universe's ongoing dance, where creation and dissolution, order and chaos, are harmonized through the flow of energy.

Consciousness as the Eternal Dance of Existence

In the grand framework of UED, consciousness emerges as the eternal dance of energy—a phenomenon that binds the universe together through perception, interaction, and evolution. It is the pulse that connects all forms of existence, from the smallest quantum particles to the vast expanses of galaxies, shaping reality through its active expression.

By reaffirming consciousness as energy's eternal dance, we uncover its profound role in defining and sustaining the cosmos. Consciousness is not just a reflection of the universe's dynamic equilibrium; it is the force that perpetuates it, ensuring that the dance of existence continues, infinite and ever-changing. Through consciousness, the universe breathes, adapts, and evolves, a timeless phenomenon that encapsulates the beauty and complexity of all that is.

Chapter 15: Life: An Expression of Energy Dynamics

As we journey deeper into the principles of Unified Energy Dynamics (UED), the emergence of life stands as one of the most extraordinary manifestations of energy at work. Building on the understanding of consciousness as energy's active expression, life represents a complex and dynamic interplay where energy transitions create systems capable of sustaining themselves, adapting, and evolving.

Life, in its myriad forms, exemplifies how energy flows through cycles, creating localized pockets of stability and activity within the universal energy medium. From the cellular mechanisms powering the smallest organisms to the intricate ecosystems connecting all living beings, life demonstrates the profound ability of energy to organize, interact, and innovate. This chapter explores life as a dynamic expression of energy, focusing on its foundation as biological energy loops, its evolution driven by energy transitions, and its inseparable connection to the universal energy fabric.

Through this lens, we uncover how life is not merely a phenomenon confined to biology but a reflection of the broader energy dynamics shaping the cosmos.

15.1 Introduction: Life as Energy in Motion

Life is one of the most intricate and remarkable manifestations of energy principles in the universe. It represents a continuous dance of stability and dynamism, where energy flows, transforms, and sustains complex systems capable of growth, adaptation, and evolution. Within the framework of Unified Energy Dynamics (UED), life emerges as a dynamic interplay of energy loops that connect biological systems to the universal energy medium.

At its core, life is energy in motion—a phenomenon that transforms raw potential into active processes, creating systems that maintain themselves and interact with their surroundings. From the cellular machinery driving metabolic processes to the intricate networks of ecosystems, biological systems embody the principles of energy transitions and flow.

This introduction sets the stage for exploring the unique role of life within UED. Biological systems are not merely isolated entities; they are localized expressions of the universal energy framework. By examining life through the lens of UED, we gain a deeper understanding of how energy shapes not only individual organisms but also the interconnected web of life that sustains the planet. Life, in this context, becomes a testament to the elegance and complexity of energy dynamics in action.

15.2 Biological Systems as Energy Loops

Life thrives on the cyclic flow of energy, where living organisms act as dynamic energy loops within the larger framework of Unified Energy Dynamics (UED). Biological systems continuously draw energy from their surroundings, transform it into usable forms, and release it back into the environment, creating a seamless cycle that sustains life. These loops are the essence of life's resilience and adaptability, showcasing how energy transitions form the foundation of biological existence.

The Cyclic Nature of Energy Flow in Living Organisms

Living organisms function as intricate energy processors, taking in raw energy, converting it, and utilizing it to sustain growth, repair, and reproduction. This cyclic energy flow is evident in every biological process, from the cellular to the ecological level.

Input, Conversion, and Output: Organisms extract energy from external sources like sunlight, organic matter, or inorganic compounds, converting it into chemical forms such as ATP (adenosine triphosphate). This energy fuels essential functions and is eventually released back into the environment as heat or waste.

Energy Efficiency: The efficiency of these cycles varies across species and ecosystems, reflecting the adaptive strategies that organisms employ to survive in changing environments.

Cellular Processes: The Foundation of Life's Energy Exchange

At the heart of every organism, cellular processes form the primary loops of energy exchange. Cells are the basic units of life, where energy transitions are meticulously managed to sustain function and vitality.

Photosynthesis and Respiration: In plants, photosynthesis captures sunlight and transforms it into chemical energy stored in glucose. Cellular respiration, in turn, releases this energy to power cellular activities. These processes exemplify the transformation of stable energy into active energy and back again.

Molecular Machines: Within cells, molecular structures like mitochondria and chloroplasts operate as energy converters, ensuring the seamless transition of energy into forms usable by the organism. These "machines" exemplify the sophistication of energy dynamics at a microscopic level.

Homeostasis and Repair: Cellular energy loops maintain homeostasis, ensuring stability amid external fluctuations. They also facilitate repair and adaptation, enabling organisms to thrive in dynamic environments.

Ecosystems: Interconnected Loops of Energy Transfer

Beyond individual organisms, ecosystems represent expansive networks of energy loops, where energy flows between species, populations, and environmental components. These interconnected loops demonstrate how life integrates with the broader energy medium.

Energy Flow in Food Chains: Energy moves through ecosystems in trophic levels, from primary producers to consumers and decomposers. Each level represents a step in the cyclic transfer of energy, ensuring that no energy is wasted.

Symbiotic Relationships: Many species engage in symbiotic energy exchanges, such as the mutualism between pollinators and flowering plants or the nutrient cycling facilitated by decomposers. These relationships highlight the collaborative nature of energy loops in ecosystems.

Global Energy Balance: Ecosystems contribute to the planet's energy balance by regulating atmospheric gases, storing carbon, and maintaining hydrological cycles. These large-scale energy loops connect life to the broader dynamics of the Earth's systems.

Life's Energy Loops in Unified Energy Dynamics

In short, biological systems as energy loops illustrate the elegance and efficiency of life's design within the UED framework. From the microcosm of cellular processes to the macrocosm of ecosystems, these loops embody the principles of energy flow, transformation, and sustainability. They not only sustain life but also highlight the interconnectedness of all living beings within the universal energy medium. Through these loops, life becomes a living testament to the dynamic harmony of energy in motion.

15.3 Evolution: A Product of Energy Transitions

Evolution, the gradual unfolding of life's diversity and complexity, is deeply rooted in the principles of Unified Energy Dynamics (UED). At its core, evolution is a dynamic process driven by energy transitions, where the interplay of stability and motion gives rise to adaptation, complexity, and the emergence of new forms of life. Through the lens of UED, evolution can be understood as an energetic phenomenon where life adapts to harness and optimize energy for survival and growth.

Energy Transitions Driving Adaptation and Complexity

The evolutionary process hinges on energy transitions, where organisms adapt to changing environments by modifying how they acquire, utilize, and store energy. These transitions fuel the development of more complex structures and behaviors, enabling life to persist and thrive.

Energy as a Selective Force: The availability of energy resources in an environment dictates which traits are favored. Organisms that can efficiently access and utilize energy have a greater chance of survival and reproduction, driving the evolutionary process.

Complexity Through Energy Adaptation: Over time, energy demands lead to the evolution of intricate systems and behaviors. For example, the development of photosynthesis allowed early organisms to harness solar energy, while the evolution of nervous systems enabled advanced coordination and response to stimuli.

Mutation and Natural Selection as Energetic Phenomena

Mutation and natural selection, the fundamental mechanisms of evolution, can be viewed through the lens of energy dynamics. They represent shifts in energy states that influence the trajectory of life.

Mutations as Energy Disruptions: Mutations arise from energetic disruptions at the molecular level, such as radiation or chemical exposure. These changes alter the genetic code, creating variations that may enhance or hinder an organism's ability to interact with its energy environment.

Natural Selection and Energy Optimization: Natural selection favors organisms that adapt to energy challenges, ensuring the perpetuation of traits that enhance energy acquisition, efficiency, and survival. This process aligns with the UED principle of balancing stability and dynamism in energy interactions.

Evolution as a Feedback Loop: Mutation introduces variability, while natural selection refines this variability based on energy dynamics. Together, they form a feedback loop where life continually adapts to optimize energy use in response to environmental changes.

The Role of Energy Efficiency in Evolutionary Pathways

Energy efficiency is a cornerstone of evolutionary success, shaping the pathways organisms take over generations. Evolution tends to favor traits and systems that maximize energy use while minimizing waste, leading to the development of optimized biological designs.

Adaptive Efficiency: Traits that improve energy efficiency—such as streamlined body shapes, efficient metabolic pathways, or cooperative behaviors—are more likely to persist in populations. For instance, the evolution of endothermy in mammals allowed for better temperature regulation, enhancing survival in diverse environments.

Specialization and Diversification: Energy efficiency drives specialization, where organisms evolve unique strategies to exploit specific niches. This diversification reduces competition and increases the overall energy efficiency of ecosystems.

Trade-offs in Evolution: Evolution often involves trade-offs between energy demands and benefits. For example, larger brains in humans require significant energy, but their advantages in problem-solving and social interaction outweighed the energetic costs.

Evolution as Energy's Creative Journey

Thus, through the lens of UED, evolution emerges as a creative journey powered by energy transitions. From the molecular scale of mutations to the grand scale of ecological adaptation, energy shapes the trajectory of life's diversity and complexity. By understanding evolution as a product of energy dynamics, we gain deeper insights into the profound interplay between life and the universal energy medium, where the drive for efficiency and adaptation fuels the ongoing saga of existence.

15.4 The Connection Between Life and Universal Energy

Life is not an isolated phenomenon but an integral part of the universal energy medium. Through the lens of Unified Energy Dynamics (UED), life forms are understood as localized expressions of the vast and interconnected energy system of the cosmos. The principles of stability and dynamism that govern energy transitions are intricately woven into the very fabric of life, sustaining its processes and connecting it to the broader universe.

Life Forms as Localized Expressions of the Universal Energy Medium

Every living organism is a microcosm of the universal energy medium, embodying its principles within a localized and dynamic form. Life transforms raw energy from its surroundings into processes that sustain and propagate existence.

Energy Localization: Life forms are clusters of organized energy, drawing from the universal pool to create structure and function. From the molecular energy in cells to the metabolic processes powering entire organisms, life captures and utilizes energy with remarkable precision.

Dynamic Adaptation: The ability of life to adapt to environmental changes reflects its connection to the broader energy medium. Organisms continuously adjust their energy acquisition and utilization strategies to align with the shifting dynamics of their surroundings.

The Interplay Between Stability and Dynamism in Sustaining Life

Life thrives on the delicate balance between stability and dynamism—two fundamental principles of UED. This interplay is evident in the processes that sustain life, enabling it to remain resilient while continuously evolving.

Stability as a Foundation: Stable energy structures, such as molecular bonds and cellular frameworks, provide the foundation for life's processes. These structures ensure continuity and order amidst constant energy exchanges.

Dynamism Driving Change: Active energy transitions, such as metabolic reactions and ecological interactions, drive the dynamism necessary for growth, reproduction, and adaptation. This dynamism ensures that life remains flexible and responsive to changes in its environment.

Balance as the Key to Survival: The equilibrium between stability and dynamism is crucial for sustaining life. Too much stability stifles growth, while excessive dynamism leads to chaos. Life's ability to navigate this balance reflects its deep integration with the universal energy medium.

Bridging the Microcosm of Life with the Macrocosm of the Universe

Life serves as a bridge between the microcosmic and macrocosmic scales of energy dynamics, linking individual organisms to the broader universe. This connection highlights the shared principles that govern both the smallest and largest expressions of energy.

Microcosmic Dynamics: At the cellular and molecular levels, life mirrors the principles of energy dynamics found in the quantum realm. Processes like energy absorption, transformation, and release echo the behaviors of particles within the universal energy medium.

Macrocosmic Integration: On a larger scale, ecosystems and biospheres connect to the planetary and cosmic energy cycles. For instance, the Earth's energy balance, driven by solar radiation and atmospheric interactions, sustains the delicate conditions necessary for life.

Universal Resonance: Life reflects the universe's ability to organize and evolve through energy. The intricate processes of living organisms resonate with the broader patterns of cosmic energy flow, revealing a profound interconnectedness.

Life as a Unified Expression of Energy

In summary, the connection between life and universal energy underscores the profound unity of existence. Life forms, as localized expressions of the energy medium, embody the same principles of stability and dynamism that govern the cosmos. By bridging the microcosmic and macrocosmic realms, life reveals its inseparable bond with the universal energy pool, highlighting the elegance and harmony of energy dynamics as they unfold across scales. Through this connection, life becomes a living testament to the continuity and creativity of the universe.

15.5 Conclusion: Life as the Universe's Dynamic Reflection

Life is a profound and dynamic reflection of the universe's energy principles, an ongoing dance of transformation, adaptation, and interconnectedness. Within the framework of Unified Energy Dynamics (UED), life embodies the seamless interplay of stability and dynamism, mirroring the cosmos's eternal rhythm of creation and evolution. From the cellular processes that power individual organisms to the ecosystems that sustain entire biospheres, life is a testament to the universe's capacity for complexity and coherence.

Life as an Ongoing Dance of Energy, Evolution, and Interconnectedness

Life is not static; it is a perpetual flow of energy, constantly evolving to meet the challenges and opportunities of its environment. This evolution is driven by the same energy transitions that govern the universe, highlighting life's role as a microcosmic expression of cosmic principles.

Energy in Motion: Life thrives on the dynamic exchange of energy, cycling through processes that sustain growth, adaptation, and reproduction. This dance of energy is mirrored in the broader cosmos, where stars, galaxies, and particles evolve through similar energetic interactions.

Interconnectedness Across Scales: The web of life is deeply interconnected, with each organism, ecosystem, and biosphere contributing to the larger energy network. This interconnectedness reflects the universal principle that no entity exists in isolation; all are part of a greater whole.

The Role of Life in Contributing to the Universal Equilibrium

Life plays a vital role in maintaining the balance of energy dynamics within the universe. By mediating the flow of energy between stability and dynamism, life contributes to the equilibrium that sustains existence.

Stabilizing Energy Cycles: Through processes like photosynthesis, respiration, and nutrient cycling, life helps regulate energy flows, maintaining the stability of ecosystems and planetary systems.

Driving Innovation and Complexity: Evolutionary processes driven by life enhance the universe's capacity for complexity and innovation, enriching the fabric of existence. Life's adaptations not only reflect energy dynamics but also influence them, creating a feedback loop of growth and change.

Harmonizing the Micro and Macro: Life acts as a bridge between the microcosm of cellular processes and the macrocosm of universal energy flow, ensuring that energy transitions are integrated across scales.

Life as the Universe's Reflection

Life is more than a product of the universe; it is a reflection of its dynamic essence. Through its cycles of energy, evolution, and interconnectedness, life exemplifies the principles that govern the cosmos. It is a living expression of the universe's creativity, adaptability, and unity.

In the grand framework of UED, life is not an isolated phenomenon but a vital component of the universal energy equilibrium. It contributes to the ongoing dance of existence, where energy flows, transforms, and sustains the harmony of the cosmos. By understanding life as the universe's dynamic reflection, we deepen our appreciation for the profound interconnectedness of all things, recognizing life's unique role in the eternal symphony of energy.

Summary of Part IV

Consciousness and Life in Unified Energy Dynamics

Part IV delves into the profound concepts of consciousness and life, examining their roles as dynamic expressions of the universe's energy principles. This section bridges the abstract and tangible, exploring how energy transitions manifest in awareness, perception, and the evolution of living systems.

Consciousness: The Bridge Between Energy States

The discussion begins with consciousness as energy in transit—a phenomenon that transcends individual awareness to integrate with the universal energy pool. Consciousness emerges as a unifying principle, linking stable and active energy states and enabling the universe to perceive, adapt, and evolve. Through themes such as quantum awareness, human intellect, and the universal fabric of consciousness, this part highlights how awareness is not limited to sentient beings but is an inherent feature of energy dynamics at every scale.

Life: The Dance of Energy and Evolution

Building on the foundation of consciousness, Part IV transitions to life as a complex and dynamic interplay of energy. Biological systems are explored as energy loops, where cellular processes and ecosystems exemplify the cyclic nature of energy flow. Evolution is reframed as a product of energy transitions, driven by adaptation, mutation, and natural selection, showcasing life's capacity to innovate and optimize energy use over generations.

The Connection Between Life and Universal Energy

Life forms are presented as localized expressions of the universal energy medium, bridging the microcosm of biological systems with the macrocosm of cosmic energy. This interconnectedness reveals how life is deeply integrated into the universal equilibrium, reflecting the same principles of stability and dynamism that govern the universe.

Concluding Insights

Part IV concludes with an affirmation of consciousness and life as integral components of the universe's dynamic equilibrium. They are portrayed as reflections of the cosmos's creative essence, emphasizing the interconnectedness of all existence through energy dynamics. By integrating these concepts, this section provides a holistic understanding of the universe as a living, evolving entity, bound together by the principles of Unified Energy Dynamics.

What If

Imagining Alternatives for Part IV

THE *What If* section invites readers to explore thought-provoking scenarios that challenge conventional ideas about consciousness and life within Unified Energy Dynamics (UED). These hypothetical questions aim to inspire curiosity, broaden perspectives, and deepen understanding of the interconnectedness of existence.

What If Consciousness is Universal, Not Individual?

Could consciousness exist as a singular, universal phenomenon rather than being fragmented across individuals? If so, might human awareness simply be localized expressions of a collective universal consciousness? This perspective could redefine how we understand individuality, shared experiences, and the interconnectedness of life.

What If Life Emerges Wherever Energy Transitions Exist?

If life arises as a product of energy transitions, could it exist in forms vastly different from those on Earth? What if alien life is not carbon-based but instead thrives on energy flows such as plasma dynamics or quantum vibrations? This scenario challenges our Earth-centric view of life and expands the possibilities for its existence in the cosmos.

What If Evolution is a Universal Principle, Not Limited to Biology?

Could the concept of evolution apply beyond living organisms, influencing systems like stars, galaxies, or even ideas and societies? If so, might the principles of adaptation and optimization inherent in energy dynamics govern the evolution of all systems, not just life on Earth?

What If Consciousness Shapes Reality More Than Perceives It?

What if consciousness actively creates reality rather than merely perceiving it? This could imply that energy transitions are influenced by the intentions and awareness of conscious entities, hinting at a universe where observation and interaction are integral to its fabric.

What If Energy Loops Are the Basis of Immortality?

Could harnessing and optimizing energy loops within biological systems allow for indefinite life? If energy flows could be perpetually sustained and cellular degradation reversed, might this unlock the potential for biological immortality or even the emergence of life forms beyond physical limitations?

What If Life's Interconnectedness is the Universe's Consciousness?

What if the interconnectedness of life systems, from cells to ecosystems, represents the universe's own self-awareness? In this view, every act of adaptation, cooperation, or evolution could be a reflection of the cosmos perceiving and understanding itself.

Purpose of the *What If* Section

These speculative scenarios encourage readers to think beyond established frameworks, exploring the possibilities that arise when energy dynamics are viewed as the foundation of both consciousness and life. By imagining these alternatives, we deepen our appreciation for the profound and interconnected dance of energy that defines existence.

Part V: Philosophical and Metaphysical Reflections

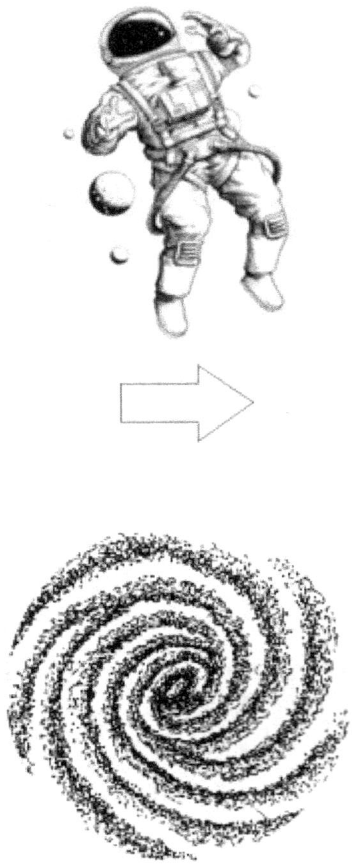

Chapter 16: Philosophical Insights: Energy as Existence

As we progress through the profound narrative of Unified Energy Dynamics (UED), the philosophical essence of energy emerges as a natural culmination of our exploration. In the preceding chapters, we delved into consciousness and life as dynamic manifestations of energy's principles, unraveling their interconnectedness with the universal energy medium. Now, we turn to the broader philosophical implications of energy itself as the foundation of existence.

Energy, in its dual forms of stability and activity, serves as both the silent essence and the dynamic force that shapes reality. It is the unifying principle that connects the tangible and intangible, the observed and the observer, offering a framework to understand the infinite universe and humanity's place within it. This chapter ventures into the deeper meanings of energy as the essence of all existence, exploring its silent stability, creative dynamism, and the profound relationship between the cosmos and human perception. Through these insights, we aim to bridge the realms of science and philosophy, revealing energy as the timeless thread weaving through the fabric of reality.

16.1 Introduction: Energy as the Core of Existence

At the heart of all existence lies energy, the fundamental force that defines reality. It is both the silent essence that underpins the universe and the dynamic expression that shapes its evolution. Within the framework of Unified Energy Dynamics (UED), energy is understood as the source of all creation and transformation, balancing stability with motion to sustain the cosmos.

Energy's dual nature—stable and active—serves as the cornerstone of existence. Stable energy provides the unchanging foundation upon which the universe rests, while active energy drives the processes of creation, transformation, and interaction. Together, they form the dynamic equilibrium that governs reality, influencing everything from the behavior of particles to the evolution of consciousness and life.

This duality extends beyond the physical realm, shaping human understanding of reality. It guides how we perceive the universe, make sense of our place within it, and explore the intricate connections between the seen and unseen. By reflecting on energy as both essence and expression, we gain a deeper appreciation for its role not only as a scientific concept but also as a philosophical principle that bridges the tangible and the intangible, the finite and the infinite.

16.2 Stable Energy: The Silent Essence of Reality

Stable energy forms the bedrock of existence, providing the unchanging foundation upon which the dynamic universe unfolds. Within Unified Energy Dynamics (UED), stable energy represents the quiet, yet omnipresent, essence of reality—a reservoir of latent potential that remains constant even as the cosmos undergoes continuous transformation.

Stability as the Foundation of Existence

In a universe defined by perpetual motion and change, stability serves as the necessary anchor that ensures coherence and order. Without stable energy, the chaos of constant activity would lack structure, making the emergence of patterns, systems, and life impossible.

A Constant Framework: Stable energy is the framework within which all dynamic processes occur. It provides the steady platform upon which motion, transformation, and interaction take place, ensuring that the universe remains balanced and functional.

Latent Potential: Stability does not imply inertia; rather, it embodies the potential for motion and transformation. This potential remains ready to activate under the right conditions, fueling the dynamic processes that drive the evolution of the cosmos.

The Unchanging Nature of Stable Energy in a Dynamic Universe

While the universe is in a constant state of flux, the stable energy medium remains unchanged, serving as the silent backdrop against which all activity occurs. This constancy ensures the continuity and harmony of existence.

Universal Stability: From the unyielding fabric of spacetime to the conserved quantities in physical laws, stable energy underpins the uniformity of the cosmos, creating a consistent reality across scales.

The Stillness in Motion: Even amidst the most chaotic transformations, stable energy remains unaffected. It is the quiet observer, allowing the universe to evolve while maintaining its foundational integrity.

Its Role as the Backdrop for All Transformations

Stable energy is not merely a passive force; it is the stage upon which the drama of the universe unfolds. It sets the conditions for dynamic processes and enables the transition from potential to reality.

A Canvas for Creation: Dynamic energy draws from the stable energy pool to drive creation and transformation. Whether in the birth of stars or the firing of neurons, stability provides the foundation for active energy to manifest.

Harmonizing Forces: Stability ensures that transformations occur within a coherent framework, preventing chaos and maintaining balance. It allows the universe to sustain itself while remaining open to infinite possibilities.

Stability as the Silent Essence

In short, Stable energy is the silent essence of reality, a force that defines the constancy underlying the universe's dynamic nature. It is both the potential for creation and the foundation of transformation, harmonizing the interplay of motion and stillness that sustains existence. By recognizing stability as an integral part of energy dynamics, we gain a deeper understanding of the balance that governs the cosmos and ensures the continuity of life and evolution.

16.3 Active Energy: The Expression of Creation and Transformation

Active energy represents the vibrant, dynamic force that drives motion, change, and transformation within the universe. It is the counterpart to stable energy, embodying the expression of potential into action. Within Unified Energy Dynamics (UED), active energy serves as the catalyst for creation, innovation, and the continuous evolution of reality.

Motion and Change as the Active Manifestations of Energy

Active energy manifests in the form of motion and change, powering the processes that shape the universe at every scale. From the oscillation of quantum particles to the expansion of galaxies, it is the force that keeps the cosmos in perpetual motion.

Energy in Action: Unlike the latent potential of stable energy, active energy is energy in motion. It drives processes such as chemical reactions, heat transfer, and the movement of celestial bodies, transforming stillness into dynamic activity.

Change as a Universal Constant: Active energy ensures that change is an inherent aspect of reality. This change fuels the evolution of matter, life, and consciousness, creating a universe that is ever-adapting and self-renewing.

The Dance Between Stability and Dynamism Shaping Reality

Reality is shaped by the delicate interplay between stability and dynamism, with active energy playing a pivotal role in maintaining this balance. This dance ensures that the universe remains both coherent and creative.

A Dynamic Harmony: Stability provides the foundation, while active energy introduces motion and transformation. Together, they create a harmonious balance where neither chaos nor stasis dominates, allowing the universe to sustain and evolve simultaneously.

Bridging Potential and Reality: Active energy acts as the bridge that transforms stable energy's latent potential into observable phenomena. It is the driving force behind every creation, transition, and interaction in the cosmos.

Creative Processes Driven by Active Energy Transitions

Active energy is the engine of creation, powering the transitions that give rise to complexity, diversity, and innovation in the universe. These processes are the essence of how energy expresses itself dynamically.

Formation of Matter and Structures: Active energy transitions drive the formation of atomic and molecular structures, shaping the physical universe. Stars, planets, and ecosystems are all products of these dynamic interactions.

Biological and Cognitive Creativity: In living systems, active energy fuels the biological processes that sustain life and the cognitive activities that drive thought and innovation. It enables organisms to grow, adapt, and create.

Cosmic Evolution: On a universal scale, active energy transitions underpin the evolution of galaxies, the birth of stars, and the emergence of complex systems. These processes reflect the universe's inherent creativity and adaptability.

Active Energy as the Pulse of Creation

Thus, active energy is the vibrant pulse of creation and transformation, the dynamic force that shapes the universe and sustains its evolution. By driving motion, change, and innovation, it transforms the latent potential of stable energy into the observable phenomena that define reality. Through the lens of UED, active energy is not only a manifestation of dynamism but also a testament to the universe's boundless capacity for creativity and renewal.

16.4 The Infinite Universe and Humanity's Place within It

The universe is an infinite and ever-evolving system of energy, a dynamic interplay of stability and transformation that shapes existence at every scale. Within this vast cosmic framework, humanity finds itself as both a participant and an observer, deeply interconnected with the universal energy medium. Unified Energy Dynamics (UED) offers a profound lens through which to understand humanity's place within this infinite system, revealing insights into existence, purpose, and the interconnectedness of all things.

The Universe as an Ever-Evolving System of Energy

The universe is a testament to the continuous flow and transformation of energy, where every phenomenon is a reflection of energy in motion. It is neither static nor finite; instead, it thrives on the dynamic interplay of creation, dissolution, and renewal.

A Dynamic Energy System: The cosmos evolves through cycles of energy transitions, from the birth and death of stars to the expansion of galaxies and the emergence of new systems. These transitions are powered by the infinite reservoir of energy that underpins reality.

Eternally Transforming: The universe's infinite nature ensures that it is always in a state of becoming, adapting to the flow of energy through its structures. This constant evolution reflects the principles of stability and dynamism inherent in UED.

Humanity's Connection to the Universal Energy Medium

Humanity, as a product of the universe, is intrinsically linked to the energy medium that sustains all existence. Every aspect of human life, from biological processes to intellectual endeavors, is a localized expression of the universal energy dynamics.

Energy Within and Beyond: Human beings are composed of the same energy that forms stars, planets, and galaxies. This connection highlights humanity's place as a microcosm of the universe, bound by the same principles that govern the cosmos.

Conscious Participants: Unlike inanimate matter, humanity possesses consciousness, enabling it to perceive, interact with, and even influence the universal energy flow. This awareness places humans in a unique position to explore and reflect on their relationship with the cosmos.

Interdependence and Impact: Humanity's actions, whether through innovation, exploration, or environmental interaction, contribute to the energy dynamics of the planet and beyond. This interdependence underscores the shared destiny of humanity and the universe.

Philosophical Reflections on Existence, Purpose, and Interconnectedness

The infinite nature of the universe invites profound philosophical contemplation about humanity's existence and purpose within it. Through the lens of UED, these reflections emphasize interconnectedness as the essence of being.

Existence as Energy Expression: Humanity's existence is a dynamic expression of the universal energy medium, a testament to the cosmos's ability to create complexity and consciousness. This perspective shifts the focus from individuality to a shared origin with all existence.

Purpose in Interconnectedness: Humanity's purpose can be seen as fostering harmony within the universal energy flow, aligning actions with the principles of stability and dynamism. By recognizing this interconnectedness, humans gain a deeper sense of responsibility toward themselves and the cosmos.

Unity Through Diversity: The infinite diversity of the universe reflects its fundamental unity. Humanity, as part of this diversity, contributes its unique awareness and creativity to the evolving tapestry of existence.

Humanity in the Infinite Dance of Energy

In summary, the infinite universe is a boundless system of energy transitions, where humanity occupies a unique and interconnected place. By understanding its connection to the universal energy medium, humanity gains a deeper appreciation of its role as both a reflection and a participant in the cosmos. Through philosophical insights into existence and purpose, Unified Energy Dynamics invites us to embrace our interconnectedness with the infinite dance of energy, recognizing the profound unity that binds all things in the vast expanse of reality.

16.5 Conclusion: Energy as the Philosophical Bridge between Reality and Perception

Energy, in its dual forms of stability and activity, serves as the philosophical bridge that connects reality to perception. It is the fundamental principle that underpins the fabric of existence, shaping both the tangible world we observe and the intangible awareness that allows us to perceive it. Within the framework of Unified Energy Dynamics (UED), the interplay between stable and active energy reflects the profound harmony that sustains the universe and defines our understanding of it.

The Harmony of Stable and Active Energy as a Reflection of Existence

The universe thrives on the balance between stable and active energy, a harmony that manifests across all scales of existence. Stability provides the foundation, the silent essence that ensures coherence, while active energy drives the motion and transformation that bring reality to life.

A Unified Dance: The seamless interplay of stability and activity reveals the universe as a cohesive system, where creation and dissolution, order and change, coexist in dynamic equilibrium. This balance is not just a feature of the cosmos but a reflection of its very essence.

Reality Shaped by Energy: All phenomena, from the motion of galaxies to the thoughts in a human mind, emerge from the interaction of stable and active energy. This dynamic equilibrium ensures the continuity and adaptability of existence, providing the framework for all that is.

Insights Into Humanity's Role in the Infinite Dance of Energy

Humanity occupies a unique place in the universal dance of energy, possessing both the capacity to perceive it and the agency to influence it. This dual role invites philosophical reflection on the purpose and responsibility of human existence.

Conscious Observers and Participants: Unlike inanimate matter, humans are conscious participants in the energy dynamics of the universe. This awareness allows us to reflect on our interconnectedness with the cosmos and recognize our shared essence with all forms of existence.

Catalysts of Harmony: Humanity's actions, guided by an understanding of energy dynamics, have the potential to align with the universal equilibrium. By fostering harmony between stability and dynamism, humans contribute to the ongoing evolution of the cosmos.

A Bridge Between Scales: As microcosms of the universe, humans embody the same principles of energy dynamics that govern the cosmos. This unique position allows us to explore, interpret, and influence the infinite dance of energy, bridging the tangible and the intangible, the finite and the infinite.

Embracing the Philosophical Unity of Energy

Energy is more than a scientific concept; it is the essence of existence and the bridge between reality and perception. Its harmony as stable and active forms reflects the interconnectedness of all things, offering insights into the nature of the universe and humanity's place within it.

Unified Energy Dynamics reveals humanity as both a reflection and a participant in the cosmic dance of energy. By embracing our role within this infinite system, we gain not only a deeper understanding of existence but also a greater sense of purpose and responsibility in shaping the harmony of the universe. Through this perspective, energy becomes a timeless principle that unites science, philosophy, and the essence of life itself.

Chapter 17: The Boundless Cosmos: Beyond the Observable

As we traverse the narrative of Unified Energy Dynamics (UED), our understanding of the universe continues to expand, moving from the observable to the boundless. In previous chapters, we explored energy as the essence of existence, the foundation of life, and the bridge between reality and perception. Now, we shift our focus to the cosmos itself—a vast, dynamic system that extends far beyond what can be seen or measured.

The observable universe, with its galaxies, stars, and planets, is only a fraction of the cosmic expanse. Beyond it lies an infinite realm of energy, potential, and phenomena that challenge our understanding of reality. Through the lens of UED, this chapter revisits the nature of the universe, reevaluates concepts such as dark matter, and embraces the limitless possibilities of energy as the ultimate framework for existence. By venturing beyond the observable, we uncover new perspectives on the cosmos and humanity's quest to comprehend the infinite.

17.1 Introduction: Expanding Horizons beyond the Observable

The universe, as we perceive it, is an intricate tapestry of stars, galaxies, and cosmic phenomena observable through telescopes and instruments. Yet, this visible dimension represents only a small fraction of the cosmos. Beyond what we can see lies a vast, enigmatic realm that challenges conventional understanding and invites deeper exploration.

Unified Energy Dynamics (UED) offers a framework to move beyond the observable, enabling us to interpret the unseen dimensions of the universe through the principles of energy. By focusing on the interplay between stable and active energy, UED transcends the limits of observation, revealing the hidden scaffolding that supports the cosmic expanse. It invites us to reconsider the nature of the universe—not as a finite structure confined to what is visible, but as a boundless, evolving system of energy and potential.

This chapter embarks on a journey to expand our horizons, revisiting the nature of the universe through UED, challenging long-held concepts like dark matter, and embracing the infinite possibilities that energy offers. By venturing into the unseen, we deepen our understanding of the cosmos and our place within its boundless expanse.

17.2 Revisiting the Nature of the Universe Through UED

The universe, when viewed through the lens of Unified Energy Dynamics (UED), emerges as an energy-driven system that transcends physical boundaries. While traditional approaches often focus on observable matter and forces, UED broadens the scope, recognizing the cosmos as a boundless interplay of energy in its stable and dynamic forms. This perspective reshapes our understanding of the universe, bridging the gap between the seen and the unseen.

The Universe as an Energy-Driven System Beyond Physical Boundaries

In the framework of UED, the universe is not constrained by the limits of physical observation. It is a dynamic energy system, where stable energy forms the silent foundation, and active energy drives motion, creation, and transformation.

Beyond the Observable: The majority of the universe's components, such as dark matter and dark energy, remain unseen. UED posits that these phenomena may not be "dark" in the traditional sense but are instead manifestations of energy that exist beyond current observational capabilities.

Energy as the Essence of Existence: By focusing on energy rather than matter, UED offers a more holistic understanding of the cosmos, one that integrates physical and metaphysical dimensions into a unified framework.

Understanding the Cosmos Through Stability and Dynamism

The universe thrives on the balance between stability and dynamism, a principle that governs everything from the smallest particles to the largest galactic structures.

Stability as the Framework: Stable energy provides the foundation for the universe's coherence, ensuring that structures, from atoms to galaxies, remain intact over time.

Dynamism as the Driver: Active energy fuels the processes of change and evolution, driving the formation of stars, the expansion of galaxies, and the emergence of complexity throughout the cosmos.

Cosmic Equilibrium: The interplay of these two forces ensures that the universe is both stable enough to exist and dynamic enough to evolve, maintaining a delicate balance that sustains its boundless nature.

The Interconnectedness of Observable and Unobservable Phenomena

The cosmos is a continuum of energy where the observable and unobservable are intrinsically linked, forming a seamless whole.

Hidden Energy Systems: Phenomena like gravitational anomalies, attributed to dark matter, may reflect the interconnectedness of energy across dimensions that transcend human perception.

Observable Evidence of the Unseen: The behavior of galaxies, cosmic radiation, and large-scale structures provides indirect evidence of unseen energy dynamics, bridging the visible and invisible aspects of the universe.

A Unified Perspective: UED unites observable and unobservable phenomena into a cohesive framework, highlighting how they interact to shape the universe as an energy-driven entity.

A New Paradigm for the Cosmos

Revisiting the nature of the universe through UED reveals it as an intricate and interconnected energy system, far beyond the limits of physical observation. By embracing stability and dynamism as its guiding principles, UED offers a profound understanding of how the cosmos operates as a unified whole. This perspective not only expands our view of the universe but also deepens our connection to its infinite mysteries, bridging the gap between the known and the unknown.

17.3 Challenging the Concept of Dark Matter

Dark matter has long been a cornerstone of modern astrophysics, proposed as an invisible form of matter that constitutes a significant portion of the universe's mass. Yet, despite its theoretical importance, direct evidence for dark matter remains elusive. Unified Energy Dynamics (UED) offers an alternative perspective, challenging the conventional understanding of dark matter by reframing it through the lens of energy dynamics. This approach shifts the focus from unseen matter to the hidden energy scaffolding that supports the cosmos.

Revisiting the Dark Matter Hypothesis Through the Lens of UED

The concept of dark matter arises from observations that the visible mass of galaxies and other cosmic structures cannot account for their gravitational behavior. UED reinterprets these observations, suggesting that what is perceived as "dark matter" might instead be an aspect of energy operating beyond the boundaries of current detection methods.

Energy, Not Matter: UED posits that the gravitational effects attributed to dark matter could stem from stable energy structures that exist outside the observable spectrum. These energy forms, while invisible to conventional detection, might influence matter through their intrinsic properties.

A Continuum of Energy: Rather than existing as discrete particles, the phenomena associated with dark matter may represent the continuum of stable energy dynamics, extending into realms yet to be explored by human understanding.

Alternative Perspectives: Energy as the Hidden Scaffolding of the Cosmos

Through the UED framework, the unseen forces attributed to dark matter can be understood as manifestations of the universal energy medium, providing the hidden scaffolding that supports the observable universe.

Stability as the Cosmic Framework: Stable energy, in its silent essence, may act as the invisible structure that binds galaxies together and shapes cosmic evolution. This energy might operate on dimensions or frequencies that lie beyond human perception.

Dynamic Interactions: The interplay between stable and active energy could explain the gravitational effects seen in galactic rotation curves and large-scale cosmic structures without requiring the existence of dark matter as a separate entity.

Expanding the Energy Paradigm: By shifting the focus from matter to energy, UED opens the door to exploring phenomena that challenge traditional models, embracing the possibility that dark matter is an energetic rather than material phenomenon.

Implications of Redefining Dark Matter in Energy-Centric Terms

Redefining dark matter as a manifestation of energy rather than matter has profound implications for our understanding of the universe and its underlying principles.

A Unified Framework: Incorporating dark matter into the UED framework simplifies the cosmic model, integrating it into a cohesive understanding of stability and dynamism. This approach eliminates the need for hypothetical particles and emphasizes the universal nature of energy.

New Avenues for Exploration: Viewing dark matter as stable energy encourages the development of innovative detection methods that focus on energy dynamics rather than particle interactions. This paradigm shift could lead to breakthroughs in understanding the unseen forces of the cosmos.

A Broader Cosmic Perspective: Reframing dark matter in terms of energy dynamics expands our view of the universe, emphasizing its interconnected and energy-driven nature. It reinforces the idea that the cosmos is not a collection of isolated phenomena but a unified system governed by the principles of UED.

Rethinking the Invisible Forces of the Cosmos

Thus, challenging the concept of dark matter through the lens of UED transforms our understanding of the unseen forces shaping the universe. By interpreting dark matter as an expression of stable energy, we move beyond the limitations of traditional models, embracing a more holistic and energy-centric perspective of the cosmos. This shift not only deepens our appreciation for the boundless complexity of the universe but also inspires new questions and explorations into the infinite possibilities of energy.

17.4 Embracing the Infinite Possibilities of Energy

Energy, as the fundamental essence of the universe, offers a framework for understanding the infinite possibilities of cosmic phenomena. Within the lens of Unified Energy Dynamics (UED), energy is not merely a measurable force; it is the foundation upon which the observable and unobservable dimensions of the cosmos are built. By embracing the boundless potential of energy,

UED invites us to expand our imagination and explore realms that lie beyond current human comprehension.

Energy as the Ultimate Framework for Infinite Cosmic Phenomena

Energy governs every aspect of existence, from the smallest particles to the vast structures of galaxies. Its stability and dynamism provide a universal framework for phenomena that transcend traditional scientific boundaries.

A Universal Language: Energy is the common thread that connects all phenomena, bridging the known and unknown. From gravitational forces to electromagnetic fields, energy operates as the underlying principle that shapes the cosmos.

Infinite Manifestations: The versatility of energy allows it to manifest in countless forms, many of which remain undiscovered. UED positions energy as the ultimate framework capable of accommodating the complexities and mysteries of the universe.

The Potential for Unknown Forms of Energy Beyond Human Comprehension

Despite humanity's advances in understanding energy, the universe's vastness suggests the existence of forms and dimensions of energy that remain beyond our grasp. UED opens the door to these possibilities, encouraging an open-minded approach to exploration.

Beyond the Observable Spectrum: Just as the electromagnetic spectrum extends beyond visible light, energy likely exists in forms undetectable by current technology. These unknown forms could hold the key to understanding phenomena like dark energy, cosmic expansion, or quantum entanglement.

Unseen Dimensions of Stability and Dynamism: Energy may operate in dimensions or scales inaccessible to human perception, offering insights into the fundamental nature of reality that challenge our existing paradigms.

A Catalyst for Discovery: The recognition of unknown energy forms inspires the development of new tools, theories, and methodologies aimed at uncovering these hidden dimensions of the cosmos.

How UED Inspires Exploration and Acceptance of Cosmic Possibilities

Unified Energy Dynamics provides a philosophical and scientific foundation for embracing the infinite potential of energy. By focusing on the interplay between stability and dynamism, UED encourages exploration and fosters acceptance of the vast, untapped possibilities within the universe.

A Dynamic Perspective: UED shifts the focus from rigid models to a flexible framework that accommodates the unknown. This adaptability allows for the integration of new discoveries into an ever-evolving understanding of the cosmos.

Exploration Through Integration: By uniting observable phenomena with theoretical insights, UED bridges the gap between what is known and what is possible, fostering a holistic approach to cosmic exploration.

A Spirit of Openness: UED emphasizes the importance of curiosity and acceptance, acknowledging that the universe's infinite complexity may never be fully comprehended but is always worth exploring.

The Boundless Potential of Energy

Thus, Energy, as understood through UED, is more than a scientific principle—it is a gateway to infinite possibilities. By embracing the vast and unknown dimensions of energy, we expand our understanding of the cosmos and our place within it. UED inspires a spirit of exploration, curiosity, and acceptance, urging humanity to continue seeking answers while appreciating the profound mystery of the universe. Through this perspective, energy becomes not only the foundation of existence but also the inspiration for an endless journey of discovery.

17.5 Conclusion: The Cosmos as an Ever-Unfolding Mystery

The cosmos, as revealed through the lens of Unified Energy Dynamics (UED), is far more than the sum of its observable components. It is an ever-unfolding mystery, a boundless system of energy transitions that transcends human perception. Beyond the galaxies, stars, and planets we can see lies an infinite expanse of energy, shaping the universe in ways that challenge and expand our understanding.

Beyond the Observable: The Boundless Energy of the Universe

The universe is a dynamic entity, where visible matter and forces represent only a fraction of its true nature. Beyond the observable lies an intricate interplay of stable and active energy, forming the hidden scaffolding of existence.

Unseen Dimensions: Energy operates in realms and forms beyond human comprehension, offering infinite possibilities for phenomena that stretch the limits of our imagination. These unseen dimensions invite us to rethink our understanding of space, time, and reality itself.

Infinite Potential: The cosmos is a continuum of energy, constantly evolving and adapting. This boundless energy drives the formation of galaxies, the expansion of space, and the emergence of life, underscoring the infinite possibilities inherent in the universe.

Humanity's Role in Exploring and Understanding Infinite Possibilities

As conscious participants in the universe, humanity holds a unique position in the cosmic narrative. Our capacity for exploration, curiosity, and reflection makes us active contributors to the unfolding story of energy and existence.

Explorers of the Unknown: Humanity's quest to understand the cosmos extends beyond observation to theoretical exploration. By developing tools, theories, and frameworks like UED, we strive to bridge the gap between the known and the infinite.

Stewards of Awareness: As beings of consciousness, humans have a responsibility to explore the universe with humility and purpose. This role involves not only seeking knowledge but also fostering a deeper connection to the infinite energy that sustains existence.

A Shared Journey: Humanity's exploration of the cosmos is part of a larger, interconnected narrative, where understanding the universe contributes to our understanding of ourselves and our place within it.

A Call to Curiosity and Wonder

The cosmos, as an ever-unfolding mystery, reminds us of the vastness of existence and the boundless possibilities of energy. Through the principles of UED, we gain a deeper appreciation for the interconnectedness of all things and the infinite dimensions waiting to be explored.

Humanity's role is not merely to observe but to actively engage with the cosmos, embracing its mysteries with curiosity, humility, and wonder. In this boundless expanse of energy, we find not only the essence of the universe but also the inspiration to continue our eternal journey of discovery and understanding.

Summary of Part V

Philosophical and Metaphysical Reflections

Part V of *Unified Energy Dynamics (UED)* delves into the philosophical and metaphysical dimensions of energy as the foundation of existence. This section bridges scientific inquiry with reflective thought, exploring how the principles of UED illuminate the nature of reality, the universe, and humanity's place within the infinite cosmos.

Chapter 15: Philosophical Insights: Energy as Existence

Stable Energy as the Silent Essence of Reality: Stable energy forms the unchanging foundation of existence, providing the backdrop against which all motion and transformation occur. It embodies the latent potential that enables coherence and continuity across the universe.

Active Energy as the Expression of Creation and Transformation: Active energy drives the dynamic processes of creation, motion, and change, bringing stability into action and fostering evolution. This interplay creates the vibrancy and adaptability that define the cosmos.

The Infinite Universe and Humanity's Place Within It: The universe is portrayed as an ever-evolving system of energy, where humanity is deeply interconnected with the energy medium. This chapter reflects on the philosophical implications of this relationship, emphasizing unity and purpose within the cosmic whole.

Chapter 16: The Boundless Cosmos: Beyond the Observable

Revisiting the Nature of the Universe Through UED: The concept of the energy medium redefines the universe as an interconnected, energy-driven system that transcends physical boundaries. This perspective bridges the observable and unobservable dimensions of reality.

Challenging the Concept of Dark Matter: UED reframes the idea of dark matter as a manifestation of stable energy within the energy medium, suggesting an alternative understanding of gravitational phenomena and cosmic structure.

Embracing the Infinite Possibilities of Energy: The infinite energy pool serves as a gateway to unknown phenomena, encouraging exploration beyond conventional limits. This chapter invites readers to embrace the boundless potential of energy, inspiring curiosity and openness to the mysteries of the cosmos.

Key Themes in Part V

Energy as the Essence of Existence: Part V explores the duality of energy as both stable and dynamic, emphasizing its role as the unifying principle that governs all phenomena.

The Infinite Universe: The chapters highlight the universe's boundlessness, urging readers to expand their understanding of existence beyond the observable and embrace the possibilities inherent in the energy medium.

Humanity's Connection to the Cosmos: The reflections underscore humanity's role as participants in the universal energy system, fostering a sense of unity and responsibility within the infinite energy pool.

Part V in Context

Part V provides a philosophical and metaphysical exploration of UED, reinforcing its principles while inspiring a deeper appreciation for the interconnectedness of existence. By merging scientific insights with reflective thought, this section bridges the tangible and intangible, offering a holistic view of the universe and humanity's place within it.

Philosophical Parallels

Reflections from the Gita and Other Texts

Unified Energy Dynamics (UED) offers a profound exploration of the universe through the principles of energy. This perspective resonates deeply with the timeless wisdom of spiritual and philosophical texts such as the *Bhagavad Gita*, the *Tao Te Ching*, and the *Upanishads*. These teachings provide a philosophical lens that parallels and enriches UED's scientific narrative, bridging the realms of observation, introspection, and universal truth.

Part I: Energy as the Foundation of Reality

Bhagavad Gita (Chapter 2: Sankhya Yoga): The *Gita* describes the *Atma* (soul) as eternal, indestructible, and pervasive, akin to UED's principle of energy's constant presence and transformation. Just as energy cannot be created or destroyed, the soul transcends birth and death, reflecting the universal constancy of existence.

Tao Te Ching: Lao Tzu's notion of the Tao as the source of all things mirrors UED's portrayal of the energy medium as the underlying essence of reality. The Tao, like energy, is unnameable and infinite, yet it manifests in all forms and phenomena.

Vedanta's *Brahman*: The energy medium in UED finds a striking parallel in the concept of *Brahman*, the ultimate reality that is formless, eternal, and the substratum of all existence.

Part II: Dynamics of Stability and Motion

Bhagavad Gita (Chapter 3: Karma Yoga): The balance between stability and motion in UED reflects *Karma Yoga's* emphasis on action without attachment to results. Stability represents the inner detachment, while motion reflects the ongoing performance of duty in harmony with the universe.

Yin-Yang Philosophy: The interplay of stability and motion echoes the *yin-yang* principle, where opposing forces coexist and complement each other to create a dynamic equilibrium. Similarly, UED highlights how stability anchors reality while motion drives transformation.

Buddhism's *Anicca* (Impermanence): The constant motion of active energy in UED parallels the Buddhist teaching of impermanence, where change is fundamental to existence, and nothing remains static.

Part III: Consciousness as Energy in Transit

Bhagavad Gita (Chapter 6: Dhyana Yoga): Consciousness in UED, acting as the bridge between stable and active energy, aligns with the *Gita's* teachings on meditation, where the mind connects the physical body to higher awareness. Meditation itself reflects the dynamic energy of thought and stability of focus.

Mandukya Upanishad: The *Upanishad's* depiction of consciousness through the waking, dreaming, and deep sleep states parallels UED's view of consciousness as a spectrum of energy interactions. Both frameworks highlight the omnipresence of awareness.

Quantum Awareness and Vasudeva Sarvam Iti (Gita): UED's quantum awareness, where particles exhibit subtle forms of responsiveness, mirrors the *Gita's* assertion that all of existence is imbued with divine consciousness, reflecting the unity of all things.

Part IV: Life as an Expression of Energy Dynamics

Bhagavad Gita (Chapter 13: Kshetra-Kshetrajna Vibhaga Yoga): The *Gita* describes the body (*Kshetra*) as the field and the self (*Kshetrajna*) as the knower of the field. This resonates with UED's depiction of life as localized energy expressions within the universal energy medium, where awareness interacts with matter.

Tao Te Ching: Lao Tzu's portrayal of life as a harmonious flow of the Tao aligns with UED's understanding of life as energy loops, interconnected with the universal medium and reflective of stability and dynamism.

Evolution and Parivartanam Sarvam Jeevanam (Change is Life): UED's framing of evolution as energy transitions parallels the *Gita's* emphasis on transformation as an essential aspect of life and existence.

Unifying Themes: Energy as the Philosophical Core

These philosophical parallels emphasize that energy, as described in UED, is not just a scientific principle but also a timeless truth reflected across spiritual and philosophical traditions. The essence of UED aligns with:

Universal Oneness: The interconnectedness of energy in UED mirrors the *Advaita Vedanta* teaching of non-duality, where all existence is one undivided reality.

Dynamic Equilibrium: Stability and motion, as explored in UED, echo the balance taught in the *Tao Te Ching* and *Gita*, where harmony arises from the interplay of opposing yet complementary forces.

Infinite Potential: UED's recognition of energy's boundless possibilities reflects the open-ended, infinite nature of the universe described in both the *Gita* and the *Upanishads*.

Reflection for Readers

The philosophical parallels between UED and these ancient texts invite readers to reflect on their place within the universal energy framework. How does understanding energy as the foundation of existence deepen your connection to the cosmos? How might these timeless teachings inspire you to perceive life and consciousness as integral components of a unified reality?

By blending science and philosophy, UED not only expands our understanding of energy but also reaffirms its universal relevance, bridging the gaps between ancient wisdom and modern inquiry.

Part VI: Reflections and Applications

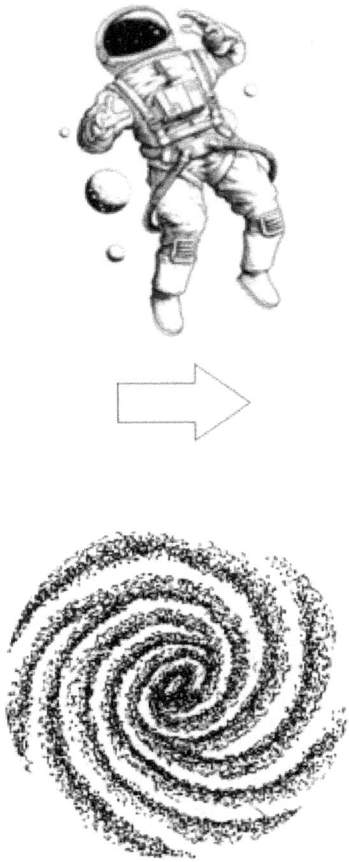

Chapter 18: Revisiting Physics Through UED

As we approach the final part of *Unified Energy Dynamics (UED)*, the journey now transitions from exploring foundational concepts to applying them in reimagining the principles that govern our understanding of the universe. In previous chapters, we delved into the essence of energy, its manifestations in life and consciousness, and its role in shaping the cosmos. Now, we turn our focus to physics—the bedrock of scientific inquiry—revisiting its core ideas through the lens of UED.

Physics, as a discipline, seeks to explain the behavior of matter, forces, and the fundamental nature of the universe. By incorporating the energy medium concept from UED, we gain new insights into established models, opening pathways to redefine critical phenomena such as light speed, cosmic distances, and nuclear reactions. This chapter bridges the theoretical elegance of UED with practical applications in physics, offering a fresh perspective on long-standing questions while inspiring innovative approaches to future exploration.

Through UED, physics evolves from a system of isolated laws to a unified framework, harmonizing the tangible and intangible, the observed and the potential. This reflection not only enhances our comprehension of the universe but also highlights the endless possibilities for discovery and innovation.

18.1 Introduction: A New Lens for Physics

Physics, as the study of the fundamental laws governing the universe, provides a framework for understanding everything from the motion of celestial bodies to the intricate behavior of subatomic particles. Yet, as our knowledge expands, so too does the need for perspectives that unify the diverse phenomena we observe. Unified Energy Dynamics (UED) offers a transformative lens for revisiting physics, integrating the concept of an all-encompassing energy medium to deepen and expand our understanding of the physical world.

The UED framework bridges the gap between classical and modern physics, creating a cohesive narrative that encompasses both the deterministic principles of Newtonian mechanics and the probabilistic nature of quantum mechanics. By positioning energy as the fundamental medium that connects all phenomena, UED provides a fresh perspective on the underlying mechanisms of gravity, electromagnetism, and other forces.

This chapter delves into how the energy medium reshapes our interpretation of foundational physics. It explores how UED not only supports existing models but also redefines key concepts like light speed, distances, and nuclear reactions. By offering a unified view of reality, UED acts as a bridge between the tangible and intangible, paving the way for innovative approaches to understanding and applying the principles of physics in the cosmic and human domains.

18.2 Supporting Existing Models with Energy Medium Concepts

Unified Energy Dynamics (UED) provides a unifying framework for revisiting foundational physics, enhancing existing models through the lens of the energy medium. By integrating stability and dynamism as core principles, UED offers a deeper understanding of classical, quantum, and relativistic phenomena while shedding new light on fundamental forces such as gravity and electromagnetism.

Revisiting Newtonian Mechanics in the Context of UED

Newtonian mechanics, the cornerstone of classical physics, describes the motion of objects through deterministic laws. Within UED, these principles gain additional depth by incorporating the energy medium as the backdrop for all interactions.

Forces as Energy Interactions: Newton's laws of motion can be reframed in terms of energy exchanges within the medium. For instance, the application of force represents the active energy required to overcome the stability of an object's inertia.

Stability and Motion: Newton's first law, the law of inertia, aligns with UED's principle of stable energy maintaining equilibrium until acted upon by active energy. This dynamic interplay reflects the balance between motion and rest within the universal energy framework.

Energy in Conservation: The conservation of momentum and energy, fundamental to Newtonian mechanics, echoes UED's assertion that energy transitions, not losses, drive the evolution of systems.

Unifying Quantum and Relativistic Phenomena with the Energy Medium

Quantum mechanics and relativity often appear as distinct and, at times, conflicting frameworks. UED bridges this gap by offering a unifying perspective through the concept of the energy medium.

Quantum Phenomena and Stability: Quantum behavior, such as superposition and entanglement, can be understood as localized disturbances within the stable energy medium. Particles exist in multiple states because they are part of an interconnected energy continuum.

Relativity and the Medium: Einstein's theory of relativity posits that space and time are interconnected and affected by mass and energy. UED complements this by suggesting that these effects arise from the medium's response to localized energy concentrations, providing a tangible basis for spacetime curvature.

Unified Scale of Interaction: UED transcends the scale divide, explaining phenomena at both macroscopic and microscopic levels as manifestations of the same underlying energy principles.

Explaining Gravity, Electromagnetism, and Other Forces Through UED Principles

The fundamental forces of nature—gravity, electromagnetism, the strong nuclear force, and the weak nuclear force—find new interpretations within the UED framework, grounded in energy medium interactions.

Gravity as a Stabilizing Energy Flow: Gravity can be seen as the medium's stabilizing response to mass, where stable energy aligns to maintain equilibrium around objects with high energy concentrations.

Electromagnetism and Energy Dynamics: Electromagnetic fields arise from the active energy transitions of charged particles. UED explains these fields as dynamic disturbances propagating through the stable energy medium, creating waves that carry both energy and information.

Nuclear Forces and Energy Stability: The strong and weak nuclear forces reflect extreme stability and dynamic transitions within atomic nuclei. UED reinterprets these forces as manifestations of the medium's interaction with highly concentrated energy states.

Enhancing Physics Through UED

Thus, by integrating the concept of an energy medium, UED enhances and unifies existing physical models, providing deeper insights into the mechanisms underlying classical and modern physics. This approach bridges the apparent divides between deterministic and probabilistic frameworks, offering a cohesive perspective on the forces and phenomena that govern the universe. Through UED, physics evolves into a more interconnected and comprehensive discipline, enriching our understanding of the cosmos.

18.3 Redefining Light Speed, Distances, and Cosmic Navigation

Unified Energy Dynamics (UED) introduces transformative insights into the concepts of light speed, cosmic distances, and navigation through the vast expanse of the universe. By incorporating the energy medium into these discussions, UED provides a dynamic perspective on how energy interactions shape the way we understand and traverse the cosmos.

Light Speed as a Dynamic Constant Influenced by the Energy Medium

In traditional physics, the speed of light is regarded as a fundamental constant, serving as the ultimate speed limit for information and matter. UED, while preserving this fundamental role, redefines light speed as a dynamic constant influenced by the characteristics of the energy medium through which it propagates.

Medium-Dependent Propagation: Light, as an electromagnetic wave, interacts with the energy medium, which could subtly affect its propagation speed based on the medium's local stability or density. This perspective echoes the behavior of light through varying materials but extends it to the universal energy framework.

Dynamic Constancy: While light speed remains consistent within a given region of the energy medium, its interactions may exhibit variations on cosmic scales where the medium's properties shift due to extreme gravitational or energetic conditions.

Expanding the Horizon: By considering light speed as part of the dynamic energy system, UED invites new exploration into phenomena like gravitational lensing, where energy medium interactions might provide additional explanatory power.

Rethinking Distances in the Context of Energy Interactions Across Space

Cosmic distances, traditionally measured in terms of light-years or parsecs, take on a nuanced meaning within the UED framework, emphasizing the role of energy interactions in defining space.

Energy-Defined Metrics: Distances in UED are not just static measurements but dynamic constructs influenced by the medium's stability and activity. Regions with heightened energy interactions, such as near black holes, could effectively "distort" perceived distances.

Interconnected Space: The energy medium connects all regions of space, suggesting that distances might be reimagined as functions of energy flow rather than static separations. This rethinking aligns with theories like quantum entanglement, where "distance" becomes less relevant.

Implications for Cosmic Observation: Rethinking distances impacts how we interpret cosmic phenomena, such as the redshift of distant galaxies, by incorporating energy medium dynamics into these measurements.

Implications of UED for Interstellar Travel and Cosmic Navigation

The energy medium framework introduced by UED opens new possibilities for interstellar travel and cosmic navigation, offering potential solutions to challenges posed by vast cosmic distances.

Harnessing Energy Medium Dynamics: Advanced propulsion systems could interact directly with the energy medium, enabling more efficient travel through space. For instance, manipulating local energy densities might reduce resistance or create pathways that shorten effective distances.

Energy-Based Navigation: Navigation systems could leverage the properties of the energy medium to map space more accurately, accounting for energy flows and transitions that influence the behavior of spacecraft and signals.

Exploring Beyond Light Speed: While traditional physics limits travel to sub-light speeds, UED hints at the possibility of bypassing these limits by exploiting the energy medium's interconnected nature. Concepts like "energy corridors" could allow for practical methods of traversing interstellar distances without violating physical laws.

A New Frontier for Cosmic Exploration

Thus, by redefining light speed, distances, and navigation through the lens of UED, we gain a more dynamic and holistic understanding of the cosmos. This perspective not only enriches our interpretation of space but also inspires innovations in how humanity explores and interacts with the universe. UED's insights bridge the gap between theoretical physics and practical applications, charting a course for a future where cosmic navigation aligns seamlessly with the boundless potential of energy dynamics.

18.4 Insights into Nuclear Reactions and Mass-Energy Interplay

Nuclear reactions, which lie at the heart of the universe's energy dynamics, provide profound insights into the interplay between mass and energy. Unified Energy Dynamics (UED) reframes these processes by highlighting the roles of stable and active energy, offering a fresh interpretation of mass-energy equivalence and the mechanisms of nuclear fusion and fission.

The Role of Stable and Active Energy in Nuclear Processes

Nuclear reactions are prime examples of the dynamic interaction between stable and active energy, where the balance between these two forms determines the release or absorption of immense energy.

Stability in Nuclear Bonds: The nuclei of atoms represent concentrated regions of stable energy, bound together by the strong nuclear force. This stability is crucial for maintaining the structure of matter and serves as a reservoir of potential energy.

Active Energy in Reaction Dynamics: In both fusion and fission, active energy disrupts the stability of nuclear bonds, triggering transformations that release or redistribute energy. These processes exemplify how active energy catalyzes significant changes in stable systems.

Energy Medium's Role: UED suggests that the energy medium itself facilitates the transfer and redistribution of energy during nuclear reactions, ensuring the conservation and continuity of energy flows.

Interpreting $E=mc^2$ Through UED: Mass as Energy in a Stable State

Einstein's iconic equation, $E=mc^2$, illustrates the profound relationship between mass and energy, asserting that mass is a concentrated form of energy. UED offers a nuanced interpretation, framing mass as energy in a highly stable state.

Mass as Compressed Stability: Within UED, mass represents energy condensed into a stable configuration, where its potential remains dormant until activated by specific processes. This perspective emphasizes the latent energy inherent in all matter.

Energy Release as Transition: During nuclear reactions, a portion of this stable energy transitions into active energy, manifesting as radiation or kinetic energy. This conversion underscores the principle that stability and dynamism are two facets of the same energy continuum.

Energy Medium Integration: UED integrates $E=mc^2$ with the energy medium concept, suggesting that the medium acts as the substrate through which energy transitions occur, maintaining equilibrium even in extreme conditions.

New Perspectives on Fusion, Fission, and Energy Transitions

Nuclear fusion and fission, the two primary processes of nuclear energy release, gain additional clarity through the UED framework.

Fusion as Energy Consolidation: In fusion, light nuclei combine to form heavier nuclei, resulting in a net release of energy as stable bonds are formed. UED frames this as an active energy-driven process that amplifies stability by releasing excess energy into the medium.

Fission as Energy Redistribution: In fission, heavy nuclei split into lighter fragments, releasing energy as the stable configuration breaks apart. UED interprets this as a redistribution of stability, where active energy drives the reorganization of matter and energy.

Energy Transitions Across Scales: UED highlights how nuclear reactions are part of a broader spectrum of energy transitions, connecting atomic-scale phenomena to cosmic processes such as star formation and supernovae.

Nuclear Reactions Through the Lens of UED

Thus, by examining nuclear reactions and mass-energy interplay through UED, we gain a deeper appreciation for the dynamic relationship between stability and activity in the universe. This perspective not only enriches our understanding of fusion, fission, and energy transitions but also emphasizes the universal principles that govern these processes. Through UED, nuclear reactions are no longer isolated phenomena but integral parts of the energy medium's boundless continuum, linking the smallest particles to the largest cosmic structures.

18.5 Conclusion: Redefining Physics Through Energy Dynamics

Unified Energy Dynamics (UED) offers a transformative lens for understanding the physical laws that govern the universe. By emphasizing energy as the foundation of all phenomena, UED enhances and reimagines traditional physics, bridging gaps between classical, quantum, and relativistic models. It introduces the energy medium as a unifying concept, providing fresh insights into foundational principles while opening pathways for innovative exploration.

How UED Enhances and Reimagines the Understanding of Physical Laws

UED reframes the core tenets of physics by integrating the interplay of stable and active energy into existing frameworks. This perspective deepens our comprehension of natural laws and provides a cohesive narrative that unites diverse phenomena.

Unifying Classical and Modern Physics: UED bridges Newtonian mechanics, quantum mechanics, and relativity by positioning energy dynamics as the universal principle underlying their apparent differences.

Redefining Fundamental Constants: Concepts like the speed of light and mass-energy equivalence are enriched by UED, which considers their relationship with the energy medium, offering a dynamic understanding of constants that govern the universe.

Recontextualizing Forces: Gravity, electromagnetism, and nuclear forces are viewed as interactions within the energy medium, providing a holistic interpretation that connects observable effects to their underlying causes.

The Potential for Future Discoveries Guided by UED Principles

The principles of UED not only reinterpret established physics but also pave the way for groundbreaking discoveries and technological advancements.

Exploring Unseen Dimensions: By highlighting the role of the energy medium in shaping reality, UED inspires the search for hidden dimensions and unknown forms of energy that may lie beyond current detection methods.

Innovations in Cosmic Navigation: UED's insights into light speed and energy interactions suggest novel approaches to interstellar travel, potentially revolutionizing humanity's exploration of the cosmos.

Sustainable Energy Solutions: Understanding mass-energy interplay through UED could lead to more efficient nuclear reactions and the development of sustainable energy systems that align with the universe's natural dynamics.

A New Era of Physics

Unified Energy Dynamics redefines physics as a comprehensive study of energy's stable and dynamic expressions across all scales of existence. It enriches our understanding of reality, transforms how we approach scientific inquiry, and inspires a future where humanity's exploration of the universe is guided by the boundless potential of energy dynamics.

Through UED, physics becomes not just a study of laws but a journey into the infinite possibilities of the cosmos, rooted in the timeless principles that connect the tangible and intangible realms of existence.

Chapter 19: Implications for Science and Technology

Unified Energy Dynamics (UED) offers a revolutionary framework for rethinking science and technology, transcending traditional boundaries to address some of humanity's most pressing challenges. From renewable energy systems to space exploration, UED's principles provide innovative pathways for sustainable advancement and deeper cosmic understanding. By grounding technological progress in the interplay of stable and active energy, UED transforms not only how we approach problem-solving but also how we envision the future of human endeavor.

This chapter explores the practical implications of UED across various domains, including energy systems, cosmology, and physics. It highlights how UED can inspire breakthroughs while addressing ethical considerations to ensure that these innovations align with humanity's values. As science and technology increasingly define the trajectory of human progress, UED offers a guiding philosophy that harmonizes innovation with the dynamic principles of the universe.

19.1 Introduction: Energy as the Pillar of Innovation

Energy has always been the driving force behind humanity's greatest scientific and technological achievements, from harnessing fire to unlocking the secrets of the atom. Unified Energy Dynamics (UED) elevates this principle, presenting energy not only as a tool but as the fundamental essence that underpins all phenomena. By redefining energy as a dynamic interplay of stability and motion within an infinite medium, UED reshapes the paradigms of science and technology, offering new pathways for innovation.

Through UED, theoretical insights are seamlessly bridged with practical applications, transforming abstract principles into actionable strategies. This approach reimagines existing systems—such as renewable energy and space exploration—while introducing novel concepts that push the boundaries of what is possible. UED inspires a vision where scientific progress aligns with the natural dynamics of the universe, creating technologies that are not only advanced but also sustainable and ethically grounded.

In this chapter, we explore how UED serves as the pillar of innovation, guiding humanity toward a future where science and technology harmonize with the principles of energy dynamics, ensuring progress that is both transformative and responsible.

19.2 Energy Dynamics in Renewable Systems and Space Exploration

Unified Energy Dynamics (UED) introduces a transformative framework for reimagining renewable energy systems and space exploration. By integrating the principles of stability and active energy, UED offers innovative approaches to enhance efficiency, sustainability, and technological advancements in these critical domains.

Revolutionizing Renewable Energy Technologies Through UED Principles

Renewable energy technologies, such as solar, wind, and geothermal systems, rely on harnessing natural energy flows. UED provides a deeper understanding of these flows, enabling more efficient and adaptable energy solutions.

Optimizing Energy Harnessing: UED's emphasis on stable and active energy interactions can inspire designs that maximize energy capture and minimize losses, such as advanced solar panels that adapt dynamically to fluctuations in sunlight.

Energy Medium Integration: The concept of the energy medium can be applied to develop systems that tap into less conventional energy sources, such as atmospheric or oceanic energy flows, by leveraging their inherent stability and motion.

Enhanced Energy Storage: Stability in UED can guide the development of high-capacity, long-duration energy storage systems, ensuring that renewable sources provide consistent power even during periods of low generation.

UED's Role in Enhancing Energy Efficiency and Sustainability

The principles of UED align naturally with the goals of sustainability, promoting energy systems that balance environmental impact with technological progress.

Reducing Waste: By understanding energy transitions as part of a dynamic equilibrium, UED can inspire systems that minimize energy waste through optimized design and efficient energy use.

Circular Energy Systems: UED's emphasis on energy loops resonates with the concept of circular systems, where waste energy is reabsorbed and reused, reducing the environmental footprint of energy production.

Global Accessibility: UED's adaptable framework encourages the creation of scalable technologies that can provide clean energy to remote or underserved regions, supporting equitable access to sustainable resources.

Advancing Space Exploration with Energy Medium Concepts

Space exploration presents unique challenges that require innovative energy solutions. UED's concept of the energy medium opens new possibilities for propulsion, navigation, and resource utilization beyond Earth.

Energy-Based Propulsion: UED principles could inspire propulsion technologies that interact directly with the energy medium, reducing reliance on traditional fuel-based systems and enabling more sustainable interstellar travel.

Dynamic Energy Mapping: The energy medium can serve as a guide for navigating cosmic distances, helping spacecraft identify energy-rich regions for refueling or support during long journeys.

Resource Utilization: By leveraging the dynamic properties of the energy medium, future missions could extract and utilize local energy resources on planets, moons, or asteroids, reducing the need for extensive supply chains from Earth.

Energy Dynamics as a Catalyst for Progress

UED offers a revolutionary perspective on energy systems, providing the tools to revolutionize renewable energy technologies and advance space exploration. By grounding innovation in the principles of energy stability and dynamism, UED enables humanity to build systems that are efficient, sustainable, and aligned with the universal dynamics of energy. This vision not only addresses current challenges but also sets the stage for a future where science and technology transcend traditional boundaries, guided by the limitless potential of energy dynamics.

19.3 How UED Transforms Our Approach to Cosmology and Physics

Unified Energy Dynamics (UED) provides a transformative lens for reinterpreting the principles of cosmology and physics, offering a more integrated and dynamic understanding of the universe. By positioning the energy medium as the foundation of all phenomena, UED reshapes how we view the cosmos, redefines established constants, and bridges the gap between experimental and theoretical physics.

Integrating UED Into Modern Cosmology: A New Understanding of the Universe

Modern cosmology seeks to explain the universe's origins, structure, and evolution. UED adds depth to this endeavor by introducing the energy medium as the unifying essence underlying observable and unobservable phenomena.

The Energy Medium as the Universe's Substratum: UED reframes the universe as a dynamic interplay of stable and active energy within an infinite medium, transcending the limitations of matter-based models. This perspective integrates dark matter and dark energy into a cohesive framework, viewing them as manifestations of energy interactions rather than separate entities.

Cosmic Evolution Through Energy Dynamics: The expansion of the universe, the formation of galaxies, and the lifecycle of stars can be understood as energy transitions within the medium, highlighting the interconnectedness of cosmic phenomena across scales.

A Unified Cosmic Framework: By connecting the microcosmic behavior of particles to the macrocosmic structure of the universe, UED offers a holistic view that bridges quantum mechanics, relativity, and cosmological observations.

Rethinking Physical Constants and Laws through the Energy Medium

Fundamental constants and physical laws form the backbone of physics, yet UED encourages us to revisit these concepts by integrating the influence of the energy medium.

Dynamic Constants: Constants such as the speed of light and Planck's constant are reinterpreted as dynamic properties influenced by the energy medium. This perspective allows for subtle variations under extreme conditions, such as near black holes or during high-energy cosmic events.

Laws of Physics as Energy Principles: UED reframes the laws of motion, thermodynamics, and electromagnetism as expressions of energy stability and dynamism. For instance, gravitational forces become stabilizing flows within the energy medium, rather than purely mass-based interactions.

Expanding the Scope of Measurement: UED opens the door to exploring energy interactions beyond current technological limits, encouraging the development of new tools to measure and understand previously inaccessible dimensions of reality.

Bridging Experimental and Theoretical Physics With UED Insights

UED serves as a bridge between experimental observations and theoretical models, creating a feedback loop that enriches both domains.

Integrating Quantum and Relativistic Models: By framing quantum behavior and relativistic phenomena within the energy medium, UED provides a unified narrative that resolves apparent contradictions between these two pillars of modern physics.

Inspiring Experimental Design: UED's principles guide the development of experiments to detect subtle energy interactions, such as mapping the energy medium's influence on particle behavior or exploring cosmic energy flows.

Theoretical Innovation: UED inspires new theories that extend beyond traditional boundaries, such as redefining the nature of spacetime as an emergent property of the energy medium's dynamics.

A New Paradigm for Physics and Cosmology

Thus, Unified Energy Dynamics transforms our approach to cosmology and physics by introducing the energy medium as the foundation of all phenomena. This perspective not only enhances our understanding of the universe but also bridges the divide between theory and experiment, enabling a more holistic and dynamic exploration of reality. Through UED, humanity gains a powerful framework to push the boundaries of discovery, redefining the fundamental principles that shape our understanding of existence.

19.4 Ethical Considerations in Advancing Energy-Based Models

The transformative potential of Unified Energy Dynamics (UED) in science and technology brings with it profound ethical responsibilities. As humanity adopts UED principles to innovate and solve global challenges, it is imperative to ensure that these advancements are aligned with values of sustainability, equity, and responsibility. This chapter addresses the critical need for ethical considerations in advancing energy-based models.

Balancing Innovation with Environmental Sustainability

Innovation must be pursued with an awareness of its impact on the environment, ensuring that technological progress aligns with the principles of ecological balance.

Sustainable Energy Practices: UED-inspired energy systems, such as renewable technologies and energy-efficient designs, must prioritize reducing environmental footprints. By leveraging UED's insights into energy loops, waste and inefficiencies can be minimized.

Mitigating Environmental Risks: Technologies rooted in UED, such as advanced nuclear energy or energy medium interactions, should incorporate safeguards to prevent unintended consequences that could disrupt ecological systems.

Long-Term Impact Assessment: Ethical frameworks should include mechanisms for evaluating the long-term sustainability of UED-driven innovations, ensuring they contribute to planetary health.

Addressing Equity and Accessibility in Energy Technologies

The benefits of UED-based advancements must be equitably distributed to ensure that all communities have access to their transformative potential.

Bridging Energy Gaps: UED technologies should be designed to address disparities in energy access, bringing affordable and sustainable solutions to underserved regions, including remote and economically disadvantaged areas.

Inclusive Innovation: Collaboration across nations, cultures, and socioeconomic groups is essential to ensure that UED applications reflect diverse needs and perspectives, fostering global inclusivity.

Preventing Technological Monopolies: Ethical governance should prevent monopolization of UED-driven technologies, ensuring that their development and deployment remain open, transparent, and beneficial to humanity as a whole.

Ethical Frameworks for Exploring and Applying UED Principles Responsibly

The exploration and application of UED principles must be guided by ethical frameworks that balance scientific curiosity with societal well-being.

Responsible Research: Scientists and technologists should adhere to principles of integrity, transparency, and accountability when exploring UED applications, particularly in sensitive areas like nuclear energy or space exploration.

Ethical Decision-Making: Policymakers and innovators must weigh the potential risks and rewards of UED-driven advancements, ensuring that short-term gains do not overshadow long-term consequences.

Global Collaboration and Regulation: International frameworks should be established to regulate the ethical use of UED technologies, fostering cooperation and preventing misuse for destructive or exploitative purposes.

Ethics as the Guiding Principle of UED Applications

Unified Energy Dynamics provides humanity with unparalleled opportunities for scientific and technological progress, but these advancements must be pursued responsibly. By balancing innovation with environmental sustainability, addressing equity and accessibility, and adhering to robust ethical frameworks, UED can inspire a future where energy-based models contribute to the well-being of all life on Earth. Ethics, as a guiding principle, ensures that the transformative power of UED is harnessed for the greater good, fostering a harmonious balance between progress and responsibility.

19.5 Conclusion: Energy Dynamics as the Future of Science and Technology

Unified Energy Dynamics (UED) represents a profound shift in how humanity understands and applies the principles of energy across science and technology. By unifying stability and dynamism as the core of all phenomena, UED provides a framework for addressing some of the most pressing challenges of our time, from sustainable energy systems to interstellar exploration. Its potential to guide humanity's scientific and technological evolution is boundless, offering pathways to innovate in ways that are both transformative and harmonious with the universe's natural dynamics.

As UED inspires groundbreaking advancements, it also emphasizes the importance of upholding ethical integrity. The pursuit of innovation must align with values of sustainability, equity, and responsibility, ensuring that technological progress benefits all of humanity without compromising the planet's future. By balancing curiosity with caution, UED-driven advancements can contribute to a world where science and technology serve as tools for creating a more connected, resilient, and equitable society.

With UED as a guiding philosophy, humanity stands at the threshold of a new era, where the principles of energy dynamics not only redefine the possibilities of innovation but also inspire a deeper appreciation for the interconnectedness of existence. As the theory continues to evolve, its impact on science and technology will not only address current challenges but also lay the foundation for a future driven by creativity, responsibility, and boundless discovery.

Conclusion

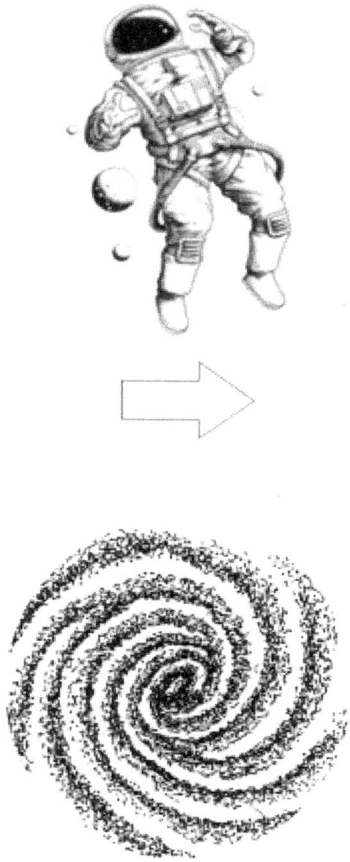

Chapter 20: Existence as Evidence: The Proof of UED

Existence itself is the ultimate testament to the principles of Unified Energy Dynamics (UED). From the rhythmic spin of subatomic particles to the majestic rotation of galaxies, the universe exhibits a harmony that defies randomness and speaks to an underlying order—an infinite energy pool that sustains and connects all things. This pool is not merely a theoretical construct but a living reality, evident in the patterns of nature, the dynamics of cosmic systems, and the emergence of consciousness.

In this chapter, we explore how the observable universe serves as proof of UED's principles, demonstrating the interplay between stability and active energy across scales. By examining the fractal patterns of spinning and orbiting phenomena, the centrality of energy hubs, and the interconnectedness of all existence, we uncover the profound intelligence inherent in nature. Existence is not just evidence of UED—it is an invitation to live in alignment with its principles, embracing the symphony of energy as both participants and witnesses to its timeless rhythm.

20.1 The Infinite Energy Pool: The Foundation of All Existence

At the heart of Unified Energy Dynamics (UED) lies the concept of the infinite energy pool—a universal constant that underpins all phenomena. This boundless medium is the origin, sustainer, and recycler of all energy and matter, serving as the foundation for existence itself. It is within this pool that stability and dynamism coalesce, creating the conditions for the cosmos to exist, evolve, and thrive.

The Energy Pool as the Universal Constant

The infinite energy pool is not a physical entity confined by space or time; it is a pervasive, self-sustaining medium that exists everywhere and governs everything. It provides the stability necessary for order while enabling the motion and transformation that define life and the universe.

A Timeless Essence: Unlike observable matter and energy, which change and dissipate, the energy pool remains constant, transcending physical limitations and acting as the ultimate source of all forms and phenomena.

Interconnected Continuum: The pool ensures that no system or entity exists in isolation. Every particle, wave, and field is connected to the energy pool, drawing from it and contributing back to its infinite reservoir.

Stability and Dynamism Unified: Within the energy pool, stable energy acts as the silent essence, while active energy drives motion, ensuring that creation, transformation, and dissolution occur in harmony.

How Existence Itself Affirms the Principles of UED

The very existence of the universe, from its smallest components to its grandest structures, offers undeniable evidence of the energy pool's presence and principles.

Patterns Across Scales: The fractal nature of the universe—seen in the similarity between atomic structures and cosmic systems—reflects the energy pool's consistent influence across scales. Electrons orbit nuclei just as planets revolve around stars, exemplifying the same underlying dynamics.

Harmony in Motion: The spinning and orbiting of particles, planets, and galaxies are not random occurrences but manifestations of energy equilibrium. These motions balance stability and dynamism, echoing the interplay within the energy pool.

Energy Recycling and Sustainability: From the transformation of mass into energy in nuclear reactions to the recycling of matter through black holes, the universe operates as a closed system sustained by the energy pool's infinite flow.

The Energy Pool as Existence's Core

The infinite energy pool is more than a theoretical concept—it is the essence of existence, the silent architect of all that is. Its principles are visible in the order, motion, and harmony that permeate the universe, affirming UED as a universal truth. By understanding the energy pool as the foundation of all existence, we gain not only a framework for explaining the cosmos but also a profound appreciation for the interconnectedness and intelligence that define reality.

20.2 Cosmic Patterns: The Signature of Harmony

The universe reveals its inherent harmony through recurring patterns, from the atomic scale to the cosmic expanse. Spinning and orbiting phenomena are not isolated occurrences but universal signatures of the energy pool's dynamic equilibrium. Unified Energy Dynamics (UED) highlights how these patterns emerge from the interplay between stable and active energy, uniting the microcosm and macrocosm in a seamless dance.

The Universality of Spinning and Orbiting Phenomena

Rotation and revolution are fundamental to the universe's structure, occurring across scales as expressions of stability and dynamism.

Spinning as a Dynamic Balancer: The rotation of particles, planets, and galaxies ensures the even distribution of energy, maintaining stability while enabling motion. Electrons spin and orbit nuclei, just as planets spin on their axes and orbit stars, reflecting a consistent principle across scales.

Orbiting as Energy Interaction: Orbital motion arises from the balance between gravitational forces and inertial dynamics in the cosmos and from electromagnetic forces at the atomic level. These orbits exemplify how stability and motion coexist in harmony, governed by the principles of the energy pool.

Synchronization Across Scales: The synchronized spinning and orbiting behaviors of systems, from quantum particles to celestial bodies, suggest a universal rhythm orchestrated by the energy medium.

How Cosmic and Atomic Structures Mirror Each Other

The structural similarities between atoms and celestial systems are striking, pointing to a fractal design inherent in the universe.

Atomic Systems as Miniature Universes: The nucleus of an atom, with its concentrated mass and positive charge, functions as the energy center, anchoring electrons in stable orbits. This mirrors how galactic centers, often housing supermassive black holes, anchor stars in orbital motion.

Energy Distribution in Patterns: Just as electrons occupy discrete energy levels around a nucleus, stars and planets organize themselves in well-defined orbits around their respective centers. These arrangements reflect the energy pool's principles of stability and dynamism at work.

Universal Intelligence in Design: The resemblance between atomic and cosmic systems underscores the universe's inherent intelligence. The same principles govern the behavior of subatomic particles and galaxies, reinforcing the interconnectedness of all existence.

Patterns as Evidence of Cosmic Harmony

Thus, the recurring patterns of spinning and orbiting phenomena reveal the profound harmony that defines the universe. Whether in the motion of electrons or the dance of galaxies, these patterns reflect the energy pool's dynamic equilibrium, uniting the tangible and intangible in a universal design. Through UED, we see that these patterns are not mere coincidences but signatures of the cosmos's inherent intelligence and interconnectedness, affirming the universal principles of energy dynamics.

20.3 Active Energy as the Lifeblood of Reality

Active energy serves as the dynamic force that drives creation, motion, and transformation in the universe. It is the lifeblood of reality, ensuring that the cosmos remains vibrant, interconnected, and ever-evolving. Unified Energy Dynamics (UED) positions active energy as the essential counterpart to stable energy, enabling the transitions and interactions that define existence. Through its influence, nothing in the universe exists in isolation; every particle, system, and phenomenon is intricately linked within the infinite energy pool.

The Role of Active Energy in Creation, Motion, and Transformation

Active energy is the catalyst for all dynamic processes, bringing stability into motion and enabling the universe to evolve and adapt.

Creation Through Energy Transitions: Active energy transforms latent potential into tangible phenomena. In the birth of stars, the ignition of nuclear fusion within stellar cores exemplifies how active energy catalyzes the creation of new structures and systems.

Motion as a Universal Dynamic: Active energy drives the spinning of particles, the rotation of planets, and the expansion of galaxies. These motions are not random but expressions of the energy pool's dynamic equilibrium, ensuring balance and coherence across scales.

Transformation Across Scales: From chemical reactions to cosmic events like supernovae, active energy facilitates the transformation of matter and energy. This continuous cycle of change sustains the universe's evolution and diversity.

Why Nothing in the Universe Exists in Isolation

The interconnectedness of all things is a fundamental principle of UED, underscored by the role of active energy in linking every system and phenomenon.

Energy Flow as a Unifying Force: Active energy flows seamlessly through the energy pool, connecting particles, fields, and systems. This flow ensures that every entity, from subatomic particles to galaxies, is part of a larger, interconnected network.

Dynamic Interdependence: The behavior of one system often influences another, reflecting the universe's inherent interconnectedness. For example, gravitational interactions between celestial bodies shape their orbits, just as electromagnetic forces bind atoms into molecules.

The Harmony of the Whole: In the energy pool, no entity exists in isolation. Even the smallest particle is dynamically linked to the cosmos, contributing to and drawing from the universal energy medium.

Active Energy as the Universe's Pulse

Thus, active energy is the pulse of the universe, driving the processes that sustain and transform reality. Its role in creation, motion, and transformation highlights the dynamic interplay that keeps the cosmos vibrant and evolving. Through its seamless flow, active energy ensures that nothing exists in isolation, reinforcing the interconnectedness and harmony of all things within the infinite energy pool. In the UED framework, active energy emerges as the lifeblood of existence, uniting the microcosm and macrocosm in a perpetual dance of creation and renewal.

20.4 Nucleus and Energy Centers: The Core of Stability

At the heart of every system, from atoms to galaxies, lies a central hub of energy—a nucleus or energy center—that anchors stability and orchestrates dynamic interactions. Unified Energy Dynamics (UED) identifies these centers as critical points where stable and active energies converge to sustain harmony and balance. They act as energy anchors, ensuring the coherence and functionality of the systems they govern.

From Atomic Nuclei to Galactic Centers: The Hubs of Energy

Energy centers are universal features, evident in both the microcosmic structures of atoms and the macrocosmic formations of galaxies.

Atomic Nuclei as Energy Hubs: In atoms, the nucleus holds the majority of mass and charge, binding protons and neutrons through the strong nuclear force. It serves as the anchor for electrons, which spin and orbit in dynamic balance, ensuring atomic stability.

Galactic Centers as Energy Anchors: At the heart of galaxies, supermassive black holes act as energy hubs, influencing the orbits of stars and interstellar matter. These centers serve as gravitational anchors, maintaining the structure and coherence of galaxies while facilitating energy redistribution through phenomena like accretion disks and jet emissions.

Fractal Consistency Across Scales: The resemblance between atomic nuclei and galactic centers highlights a fractal pattern in the universe, where energy hubs operate on consistent principles across vastly different scales.

How Active and Stable Energies Interact to Sustain Harmony

The interplay between stable and active energies at these centers is essential for maintaining the equilibrium and functionality of systems.

Stability Through Energy Anchoring: Energy centers provide a stable foundation, concentrating mass and charge to anchor surrounding components. This stability prevents chaos, ensuring that dynamic processes occur within a coherent framework.

Dynamism Through Active Interactions: Active energy emanates from these centers, driving motion and transformation. In atoms, electrons transition between energy levels, releasing or absorbing photons. In galaxies, stars and interstellar matter interact dynamically with the energy emanating from their centers.

Harmonic Balance: The constant interplay between stability and dynamism at energy centers sustains harmony. Whether binding electrons to nuclei or organizing stars around galactic cores, this balance ensures the ongoing functionality and evolution of systems.

Energy Centers as Universal Anchors

Nuclei and galactic centers exemplify the principles of UED, serving as the hubs where stable and active energies converge to sustain harmony. These energy centers anchor systems across scales, ensuring coherence and balance while enabling dynamic interactions that drive evolution and transformation. Through their interplay of stability and dynamism, energy centers highlight the interconnectedness of all systems within the infinite energy pool, reinforcing the universal principles of Unified Energy Dynamics.

20.5 Intelligence beyond Comprehension

The universe operates with an inherent intelligence that transcends human understanding, evident in its self-organizing principles and the seamless harmony of its systems. Unified Energy Dynamics (UED) identifies this intelligence as a manifestation of the infinite energy pool, where stability and dynamism interact to create complexity, coherence, and continuity. From the self-regulating patterns of nature to the profound connection between consciousness and the energy medium, this intelligence reflects the universe's intrinsic order and purpose.

Nature's Self-Organizing Principles as Proof of Universal Intelligence

The self-organization observed across scales, from atomic structures to cosmic phenomena, demonstrates the universe's innate ability to regulate and sustain itself.

Fractal Harmony: The recurring patterns of spinning and orbiting, found in both atoms and galaxies, reveal an underlying intelligence that governs motion and stability. These patterns are not arbitrary but result from the energy pool's principles of equilibrium and flow.

Dynamic Adaptation: Natural systems adapt and evolve to maintain balance, reflecting an intelligence that responds to changes within the energy medium. For example, ecosystems adjust to disruptions, while galaxies maintain structural integrity despite gravitational interactions.

Universal Coherence: The interconnectedness of all systems, from particles to celestial bodies, suggests a guiding intelligence that ensures harmony. This coherence is evident in phenomena such as the alignment of galactic spins and the synchronization of biological rhythms.

The Connection between Consciousness and the Energy Medium

Consciousness, as explored in UED, emerges as an active expression of the energy medium, bridging the tangible and intangible dimensions of existence.

Consciousness as Energy Awareness: In the UED framework, consciousness is viewed as energy's ability to perceive, interact, and adapt. It exists across scales, from the quantum awareness of particles to the reflective self-awareness of humans.

Energy Medium as the Source of Awareness: The infinite energy pool serves as the substrate from which consciousness arises. It provides the stable foundation and dynamic potential necessary for the emergence of awareness in both living and non-living systems.

Universality of Intelligence: The connection between consciousness and the energy medium suggests that intelligence is not confined to humans or biological systems. Instead, it is a universal phenomenon, present wherever energy interacts and organizes itself.

A Testament to Universal Intelligence

The self-organizing principles of nature and the connection between consciousness and the energy medium reveal a universal intelligence beyond human comprehension. This intelligence operates seamlessly through the principles of UED, ensuring harmony, coherence, and evolution across all scales of existence. By recognizing this intelligence, we gain a deeper appreciation for the interconnectedness and order of the universe, reaffirming the infinite energy pool as the source and sustainer of all that is.

20.6 The Limits of Human and Artificial Understanding

Unified Energy Dynamics (UED) challenges the limits of human and artificial understanding by revealing a universe that operates beyond conventional perception and logic. While UED provides a framework for exploring the infinite energy pool and its principles, it also highlights the inherent boundaries of our tools and intellect. This recognition is not a limitation but an invitation to embrace humility and align with the universal energy symphony.

Why UED Defines the Boundaries of Perception and Logic

Human and artificial systems of understanding are constrained by their dependence on observation, measurement, and computation, all of which operate within the confines of the energy medium.

Finite Perception in an Infinite System: Human senses and scientific instruments are limited to detecting a narrow range of phenomena, such as visible light or electromagnetic waves. The infinite energy pool, with its stable and active dynamics, extends far beyond these detectable ranges, making parts of its nature inherently inaccessible.

Dependence on Logical Constructs: Scientific models, rooted in logic and mathematics, are tools for approximating reality but cannot fully capture the infinite complexity of the energy medium. UED transcends these constructs by framing energy as both the foundation and the driver of existence, not confined to quantifiable parameters.

Artificial Intelligence Within the Energy Medium: While AI can process vast amounts of data and identify patterns, it remains a creation of human logic and is itself bound by the principles of the energy medium. AI, like humans, operates within the limits imposed by the energy dynamics it seeks to understand.

How Humanity Can Coexist With the Universal Energy Symphony

Rather than seeking to conquer or fully comprehend the energy pool, humanity's role is to coexist with its rhythms and align with its principles.

Embracing Harmony: The universe operates as a harmonious symphony, balancing stability and motion across scales. Humanity can learn from this balance, creating technologies and systems that reflect the same equilibrium, promoting sustainability and interconnectedness.

Aligning with Universal Rhythms: By recognizing the principles of UED in natural phenomena, humanity can align its actions with the universe's inherent order. This means designing systems that respect the energy medium, from renewable energy solutions to ethical exploration of cosmic and quantum realms.

Humility and Purpose: Acknowledging the limits of understanding fosters humility and a sense of purpose. Humanity's role is not to dominate the universe but to participate in its dynamic equilibrium, contributing to the ongoing dance of creation and transformation.

Coexisting With the Infinite

Unified Energy Dynamics defines the boundaries of perception and logic, reminding us that the infinite energy pool transcends human and artificial comprehension. This realization shifts our focus from control to coexistence, inviting humanity to align with the universal energy symphony. By embracing humility and harmony, we can find our place within the cosmos, participating in its timeless rhythm and contributing to the balance that sustains all existence.

20.7 Living with the Proof: Aligning with the Energy Pool

Unified Energy Dynamics (UED) provides more than a framework for understanding the universe—it offers a way of living that aligns with the infinite energy pool. The principles of UED, visible in the harmony of nature and the interconnectedness of all things, carry profound philosophical and practical implications. By recognizing existence itself as evidence of these principles, humanity is invited to embrace harmony, purpose, and coexistence within the cosmos.

The Philosophical and Practical Implications of UED

UED bridges science and philosophy, offering insights that influence not only how we understand the universe but also how we live within it.

A New Perspective on Existence: UED reframes life as a localized expression of universal energy, connected to the infinite pool. This perspective fosters a sense of unity, reminding us that we are part of a greater whole, bound by the same principles that govern stars, atoms, and galaxies.

Guiding Innovation and Sustainability: Practically, UED's emphasis on stability and dynamism encourages the development of technologies and systems that reflect these principles. From renewable energy to sustainable practices, aligning human endeavors with UED ensures long-term harmony with the environment.

Inspiration for Ethical Living: The interconnectedness revealed by UED underscores the ethical responsibility to live in balance with nature. It calls for respect, not exploitation, of the energy pool that sustains all existence.

A Call to Embrace Harmony, Purpose, and Coexistence

Living in alignment with the energy pool is not just a scientific or philosophical endeavor—it is a way of life that celebrates the beauty and intelligence of the universe.

Harmony Through Awareness: By observing the harmony of spinning planets, orbiting electrons, and evolving ecosystems, humanity can learn to mirror this balance in its actions. This harmony begins with awareness of the interconnectedness of all things and the recognition that our actions ripple through the energy medium.

Purpose Through Alignment: Understanding UED encourages us to find purpose not in domination but in contribution. Aligning with the energy pool means acting in ways that enhance its equilibrium, whether through sustainable technologies, equitable resource distribution, or the pursuit of knowledge that benefits all.

Coexistence Through Humility: The infinite energy pool is a reminder of the universe's vastness and our place within it. Accepting our limits fosters humility, inspiring us to coexist with the cosmos rather than seeking to control it. This coexistence is an invitation to participate in the eternal dance of creation and transformation.

A Life in Symphony

Living with the proof of UED is an opportunity to embrace the universe as a harmonious symphony, where every entity plays a role in maintaining balance. By aligning with the energy pool, humanity can find purpose in contributing to this harmony, fostering a life of connection, sustainability, and ethical responsibility. Unified Energy Dynamics is more than a theory; it is a call to live with awareness, humility, and reverence for the infinite intelligence that defines existence.

Appendices

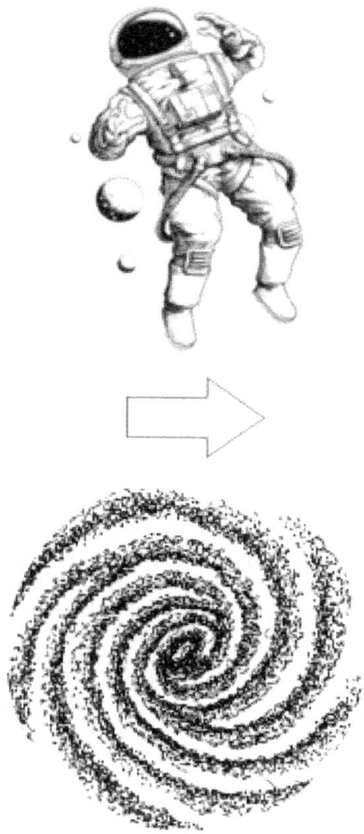

Appendix A: Advanced Insights into Unified Energy Dynamics

(Exploring Time, Photons, and the Electromagnetic Field)

UNIFIED ENERGY DYNAMICS (UED) provides a foundational framework for understanding the universe through the principles of stability and dynamism. While the main chapters focus on the energy medium and its role in shaping existence, certain advanced concepts warrant deeper exploration. This appendix expands on three critical aspects—time, photons, and the electromagnetic (EM) field—offering insights into their interconnections and their significance within the UED framework.

By examining time as an emergent property, photons as active manifestations of the EM field, and the EM field as a universal framework, we uncover layers of complexity that further affirm UED's principles. These advanced discussions not only enrich the core narrative but also invite readers to explore the profound harmony underlying all phenomena, from quantum interactions to cosmic evolution.

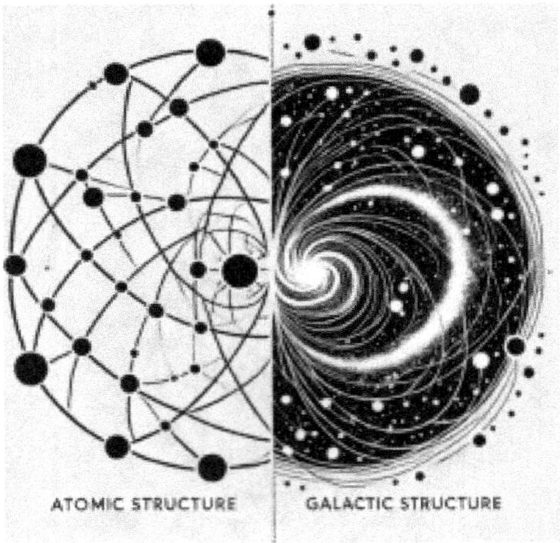

A-1

The Electromagnetic Field and the Heart's Role in UED

THE **electromagnetic (EM) field** stands as a cornerstone in the **Unified Energy Dynamics (UED)** framework, embodying the interconnected principles of stability and dynamism. Omnipresent and foundational, the EM field serves as the substrate upon which all matter and energy interactions occur. Its influence extends from the quantum scale of atomic structures to the grand designs of cosmic systems, shaping the universe in profound and measurable ways. Beyond its universal role, the EM field also finds a deeply integrated presence in the human body, particularly in the heart, which generates the strongest electromagnetic field within the body, creating a direct connection between biological systems and universal energy dynamics.

The Omnipresence of the EM Field as the Substrate of Matter and Energy

The EM field permeates the universe, existing as a constant presence even in the vacuum of space. This omnipresence makes it a fundamental framework for understanding energy dynamics.

A Timeless and Uniform Medium

The EM field exists independently of matter, providing a stable yet dynamic medium for the propagation of energy. It serves as the foundation for phenomena such as light waves, electromagnetic radiation, and particle interactions.

This stable yet flexible nature aligns seamlessly with UED's concept of energy as a dual entity—stable and active—acting as a bridge between universal equilibrium and dynamic transformations.

Latent Potential Activated Through Interactions

The field holds **potential energy** that becomes active during disturbances, manifesting as photons or other forms of electromagnetic radiation.

The heart, as a biological system, activates this latent potential at the cellular level, producing an electromagnetic field through ionic movements and rhythmic contractions. This biological EM field interacts dynamically with the stable energy medium, resonating with universal principles of energy interplay.

The Role of the EM Field in Shaping Atomic, Cosmic, and Biological Systems

The EM field is not merely a background force; it actively governs the formation, stability, and behavior of atomic, cosmic, and biological systems.

At the Atomic Scale

The structure of atoms is dictated by electromagnetic forces. Protons and electrons interact within the EM field, creating the stable configurations of atoms and molecules.

The quantized energy levels of atoms depend on the EM field's influence, enabling dynamic behavior while maintaining structural integrity.

At the Cosmic Scale

The EM field extends its influence to celestial systems, contributing to phenomena such as magnetic fields around planets and stars.

Cosmic radiation, a byproduct of high-energy processes, propagates through the EM field, acting as both an energy carrier and a window into the universe's structure and evolution.

At the Biological Scale: The Heart's Electromagnetic Field

The heart generates the strongest electromagnetic field in the human body, far exceeding that of the brain. This field, detectable several feet away, serves as a localized disturbance in the stable energy medium.

The rhythmic pulsation of the heart's EM field aligns with the body's physiological processes, creating coherence within the active energy systems of the body.

This biological EM field resonates with external fields, potentially harmonizing with the stable energy medium on a universal scale.

Evidence of the EM Field in Atomic Spectra, Material Identity, Energy Transitions, and the Heart's Dynamics

The EM field's integration into matter and energy dynamics is evident through measurable phenomena, providing tangible proof of its universal role.

Atomic Spectra as a Fingerprint

Every atom emits a unique spectrum of electromagnetic radiation, a direct consequence of electron transitions governed by the EM field. These spectra serve as a "fingerprint," identifying elements and compounds across the universe.

The quantized energy levels of atoms, shaped by the field, underscore the precision and order the EM field imparts to matter.

Material Identity Through EM Interactions

The diverse properties of materials, from conductivity to magnetism, result from the EM field's interactions with atomic structures. These interactions define the energy states and bonding characteristics of matter.

Energy Transitions and Emission

During energy transitions, such as radioactive decay or atomic collisions, the EM field mediates the release of energy as photons or other EM waves. This behavior confirms the field's role as a bridge between stable energy and active manifestations.

Heart as a Biological Bridge

The heart's EM field acts as a **localized bridge** between the stable energy medium and the body's active energy systems.

Its rhythm and coherence influence the brain, nervous system, and other organs, creating harmony within the body.

The heart's electromagnetic energy may resonate with the broader EM field, suggesting a universal connection between biological and cosmic systems.

Conclusion

The **electromagnetic field** is far more than a force—it is the universal framework that underpins the existence and behavior of matter, energy, and biological systems. From shaping atomic stability to influencing cosmic phenomena and supporting biological coherence, the EM field demonstrates the seamless integration of stable and dynamic energy within UED. Its omnipresence and measurable effects make it a vital component in understanding the interconnectedness and harmony of the universe. The heart, as a biological generator of electromagnetic energy, exemplifies the interplay between universal and localized energy dynamics, further affirming the EM field's central role in UED.

Would you like to refine this further or add more biological or cosmic examples?

A-2

Photons: The Active Form of Energy

Photons, the fundamental quanta of the electromagnetic (EM) field, embody the dynamic interplay between stability and motion within the Unified Energy Dynamics (UED) framework. As the smallest units of electromagnetic energy, photons are essential for bridging the stable energy pool with its dynamic expressions. They serve as energy messengers, facilitating interactions and maintaining the flow of energy across the universe.

Photons as the Quanta of the EM Field, Bridging Stability and Dynamism

Photons emerge from the EM field as discrete packets of energy, acting as the field's dynamic activators while preserving its inherent stability.

Energy Units in Action: Photons carry energy in quantized amounts, defined by their frequency ($E=h\nu$), where h is Planck's constant. This quantization ensures precision in energy transitions, reflecting the EM field's structured stability.

Agents of Energy Transfer: As oscillatory disturbances in the EM field, photons enable the transfer of energy across vast distances without requiring a physical medium. This capacity highlights their dual role as both stable carriers and dynamic facilitators of energy.

Stability in Motion: Despite their dynamic propagation, photons retain the stable characteristics of the energy medium, such as traveling at the constant speed of light (ccc) in vacuum. This behavior underscores their role as a bridge between the stable energy pool and its active manifestations.

Activation of Photons During Atomic Transitions and Energy Interactions

Photons are not static entities; they are activated during energy exchanges, particularly in atomic and molecular processes.

Atomic Transitions: When an electron transitions between energy levels in an atom, the difference in energy is emitted or absorbed as a photon. This process is central to phenomena like atomic spectra, where photons carry the quantized energy changes dictated by the EM field.

The activation of photons during these transitions reflects the precision of the energy medium, ensuring that only specific frequencies and wavelengths are emitted or absorbed.

Energy Interactions: Photons are also produced during high-energy interactions, such as collisions, radioactive decay, and annihilation events. These processes demonstrate the EM field's ability to transform latent energy into its active, observable form.

For example, gamma rays emitted during nuclear decay are photons carrying vast amounts of energy, showcasing the field's role in mediating extreme energy transitions.

The Role of Photons as Energy Messengers Across the Universe

Photons play a vital role in maintaining the interconnectedness and harmony of the universe by acting as efficient carriers of energy and information.

Cosmic Communication: Photons propagate through space, transmitting information about their source. For instance, the light from distant stars and galaxies reveals their composition, motion, and history through spectral analysis.

This ability to carry both energy and information makes photons indispensable for understanding cosmic phenomena.

Energy Conservation and Flow:

Photons ensure that energy transitions occur without loss, preserving the integrity of the energy medium. Their seamless propagation maintains the balance between stability and dynamism within UED.

Bridging Scales: From quantum-level interactions to cosmic-scale radiation, photons operate across all scales, unifying the principles of UED by linking the microcosm to the macrocosm.

Conclusion

Photons are the active form of energy within the electromagnetic field, serving as the dynamic expressions of the energy medium. By bridging stability and dynamism, they ensure the seamless flow of energy and information across the universe. Whether activated during atomic transitions or propagating through cosmic distances, photons exemplify the interconnectedness and precision of the UED framework, highlighting the intricate harmony underlying all existence.

A-3

Time: An Emergent Property of Energy Dynamics

Time, within the Unified Energy Dynamics (UED) framework, is not an intrinsic characteristic of the universe but an emergent property arising from energy interactions and entropy. It exists as a relational phenomenon, dependent on the dynamic processes that govern the flow of energy. Time's evolution, variability, and even cessation are intimately tied to the interplay of stable and active energy within the infinite energy pool.

Time as a Relational Property Dependent on Energy Interactions and Entropy

In the UED perspective, time emerges as a result of transitions and gradients within the energy medium.

Energy Interactions Define Time: Time is perceived and measured through the progression of energy exchanges, such as atomic vibrations, molecular changes, or macroscopic phenomena like planetary orbits. These processes rely on the dynamic flow of active energy within the stable energy medium.

Entropy as Time's Driver: The forward flow of time corresponds to the increase in entropy—a measure of disorder within a system. As energy transitions drive systems toward equilibrium, the increase in entropy provides the directional "arrow of time."

Relational and Contextual: Time is not universal but varies with local energy interactions and entropy levels. It is a contextual property that depends on the dynamics of the system being observed.

How Time Evolves with Active Energy Processes and Dissolves in the Absence of Interactions

Time's existence is sustained by the presence of active energy processes; without them, time ceases to have meaning.

Emergence Through Energy Activity: Active energy transitions create the framework for time to emerge. Processes such as chemical reactions, motion, and radiation provide the "ticks" by which time is measured and experienced.

Dissolution in Stillness: In regions devoid of interactions, where stable energy dominates without disturbances, time effectively halts. This concept aligns with the theoretical state of a perfectly homogeneous universe with no gradients or entropy changes.

Cyclic and Dynamic Nature: Time evolves as systems undergo cycles of creation and transformation, reflecting the perpetual motion of active energy. When energy interactions cease, such as at the end of a system's lifecycle, time dissolves into timeless stability.

Connections Between Gravitational Fields, Entropy, and the Flow of Time

The flow of time is influenced by gravitational fields and entropy, which alter the dynamics of the energy medium.

Gravitational Time Dilation: In strong gravitational fields, such as near massive celestial bodies, time flows more slowly. This phenomenon, predicted by general relativity, occurs because the energy medium becomes denser, altering the interaction rates that define time.

Entropy and Time's Forward Flow: In regions of high entropy, where systems approach equilibrium, the flow of time slows as fewer energy transitions occur. Conversely, in low-entropy regions, active energy processes are more pronounced, resulting in a faster perception of time.

Unified Dynamics of Time and Gravity: Gravitational fields and entropy gradients together shape the local flow of time, linking time to the structural dynamics of the universe. This interplay ensures that time is a relational property tied to energy density and interactions within the energy medium.

Conclusion

Time, as understood through UED, is an emergent property rooted in the dynamics of energy interactions and entropy. It is not an absolute entity but a relational phenomenon that evolves with active energy processes and dissolves in their absence. The connections between time, gravitational fields, and entropy underscore the profound harmony of the universe, where time flows as a natural consequence of the energy medium's perpetual motion and transformation. This perspective redefines time as a dynamic and interconnected aspect of existence, aligning seamlessly with the principles of Unified Energy Dynamics.

A-4

Interrelations: Time, Light Speed, and Entropy

Unified Energy Dynamics (UED) reveals profound interconnections between time, the speed of light, and entropy, each of which emerges from the properties of the stable energy medium. These three phenomena, often considered independently in traditional physics, are unified within UED as manifestations of energy interactions. This framework provides a cohesive understanding of how the energy medium's stability and dynamism shape the fundamental characteristics of existence.

The Speed of Light as Determined by the Stable Energy Medium's Constants

The speed of light (c) is not an arbitrary value but a direct result of the stable energy medium's intrinsic properties.

Defined by Permittivity and Permeability: The speed of light is governed by the relationship $c = 1/\sqrt{\epsilon_0 \mu_0}$, where ϵ_0 (permittivity) and μ_0 (permeability) are constants of the energy medium in a vacuum.

These constants reflect the medium's capacity to sustain electromagnetic waves, highlighting its role as a universal framework for light propagation.

Uniformity in the Stable Medium: In the stable energy medium (vacuum), ϵ0\epsilon_0ϵ0 and μ0\mu_0μ0 remain constant, ensuring that light travels at a uniform speed. This constancy reinforces the medium's timeless and homogeneous nature.

Variation in Non-Uniform Media: When light travels through regions with gravitational fields or material media, the effective speed changes due to local variations in ϵ\epsilonϵ and μ\muμ, reflecting the dynamic nature of the energy medium.

Entropy's Influence on the Properties of the Energy Medium and Its Connection to Time

Entropy, as a measure of disorder, affects the energy medium's properties and plays a critical role in the emergence and flow of time.

Energy Medium Density and Entropy: Low-entropy regions (e.g., vacuum) maintain a uniform energy medium, allowing light to travel at maximum speed and enabling well-defined time progression.

High-entropy regions (e.g., near massive objects) increase the medium's density and disorder, slowing the effective speed of light and altering time's flow.

Entropy as Time's Driver: Time flows forward as entropy increases, driven by energy transitions within the medium. This forward progression reflects the universe's natural tendency toward equilibrium.

Dynamic Interplay: As entropy shapes the medium's properties, it indirectly influences both light speed and time. This interconnectedness demonstrates the medium's adaptability and its role as the unifying substrate for these phenomena.

Unified Representation of Time, Light Speed, and Entropy Through UED Principles

Within UED, time, light speed, and entropy are interdependent aspects of the energy medium's stability and dynamism.

Time Emergence and Light Propagation: Time arises from energy interactions, while the speed of light represents the medium's ability to propagate these interactions. Together, they reflect the medium's inherent properties.

Entropy as the Unifying Factor: Entropy ties time and light speed together by influencing the medium's density and structure. High entropy slows interactions (time) and light propagation, while low entropy preserves their maximum efficiency.

Energy Medium as the Common Foundation: These phenomena are not isolated but stem from the same universal framework. The stable energy medium provides the constants and conditions that define time, regulate light speed, and guide entropy's role in cosmic evolution.

Conclusion

Time, light speed, and entropy are deeply interconnected within the UED framework, unified by the properties of the stable energy medium. The speed of light reflects the medium's constants, time emerges from its interactions, and entropy drives its evolution. Together, they reveal the profound harmony of the universe, where the interplay of stability and dynamism governs the flow of energy, information, and existence. This unified representation not only deepens our understanding but also reinforces the elegance and interconnectedness of Unified Energy Dynamics.

A-5
Philosophical Reflections

Unified Energy Dynamics (UED) invites us to view the universe through a lens of profound interconnectedness, where the electromagnetic (EM) field, photons, time, and entropy serve as both scientific phenomena and philosophical insights. These concepts not only validate the principles of UED but also highlight the intrinsic harmony and intelligence underlying existence. Together, they paint a picture of the universe as a seamless, dynamic system, governed by the infinite energy pool.

Viewing the Universe as a Vast, Interconnected EM Field

The omnipresence of the EM field within UED offers a unifying perspective of the universe as an interconnected whole.

A Universal Framework: The EM field extends across all of existence, integrating stable and active energy into a cohesive system. Its presence ensures that energy flows, matter forms, and interactions occur within a structured, harmonious medium.

The Bridge Between Scales: From the quantum behavior of photons to the cosmic dynamics of electromagnetic radiation, the EM field links the smallest particles to the largest celestial structures, revealing a fractal-like unity in nature.

Timeless and Boundless: The EM field embodies the infinite energy pool, transcending time and space while adapting to localized interactions. This timeless presence underscores the universe's inherent order and continuity.

The Implications of Photons, Time, and Entropy as Evidence of the Infinite Energy Pool

Photons, time, and entropy, when viewed through UED, emerge as profound evidence of the energy pool's role in sustaining the universe.

Photons as Energy Messengers: Photons, as quanta of the EM field, illustrate the dynamic expression of the infinite energy pool. They carry energy and information across vast distances, maintaining the universe's coherence and vibrancy.

Time as an Emergent Rhythm: Time's dependency on energy interactions and entropy reflects the dynamic adaptability of the energy pool. It emerges as a relational property, evolving with active processes and dissolving in their absence, reinforcing the energy pool's role as the substrate of existence.

Entropy as a Cosmic Driver: Entropy, which governs the flow of time and energy transitions, highlights the universe's tendency toward equilibrium. Its influence on light speed and the energy medium underscores the interconnectedness of all phenomena within the infinite energy pool.

How These Concepts Reinforce the Interconnectedness and Harmony Proposed by UED

The integration of photons, time, and entropy into the UED framework reveals a universe that operates with a profound sense of harmony and interconnectedness.

Unified Principles: These concepts demonstrate how seemingly distinct phenomena are manifestations of the same energy medium. They validate UED's assertion that the universe is not fragmented but inherently unified.

A Dynamic Balance: The interplay of stability and dynamism, evident in the behavior of photons, the emergence of time, and the evolution of entropy, highlights the universe's ability to sustain balance while fostering transformation.

An Intelligent Design: The self-regulating principles of the energy pool, evident in the precision of atomic spectra, the propagation of light, and the flow of time, point to an underlying intelligence that governs existence. This intelligence reinforces the harmony proposed by UED.

Conclusion

Viewing the universe through the lens of UED reveals a vast, interconnected system where photons, time, and entropy serve as evidence of the infinite energy pool's sustaining role. These concepts not only affirm the scientific principles of UED but also inspire a philosophical understanding of the universe as a harmonious and intelligent whole. By embracing this perspective, we deepen our appreciation for the elegance and unity of existence, as envisioned through Unified Energy Dynamics.

A-6
Implications for Future Exploration

Unified Energy Dynamics (UED) offers a transformative framework with far-reaching implications for science, technology, and philosophy. By leveraging its principles, we can pursue experimental validations and innovative applications in fields ranging from quantum mechanics to cosmology and space exploration. These efforts not only validate the theoretical foundations of UED but also open new avenues for addressing some of humanity's greatest challenges.

Potential Areas of Experimental Validation

UED's insights into photons, the electromagnetic (EM) field, and time provide fertile ground for experimental exploration, offering opportunities to deepen our understanding of energy dynamics.

Photon Behavior in Varying Energy Media: Investigating how photons propagate through different energy media, including strong gravitational fields or high-entropy systems, could provide direct evidence of the energy medium's properties.

Experiments could measure variations in light speed and photon energy in controlled environments, such as vacuum chambers with artificial gravitational gradients.

Time and Entropy Relationships: Experiments on the relationship between time flow and entropy changes in localized systems could reveal how time emerges and evolves with energy interactions.

Observing entropy-driven changes in quantum systems might shed light on time's connection to energy transitions.

EM Field Interaction Studies: Advanced spectroscopy techniques could explore how the EM field mediates energy exchanges during atomic and molecular transitions, providing deeper insights into its omnipresent role.

Applications in Cosmology, Renewable Energy, and Space Exploration

The principles of UED offer innovative pathways for addressing practical challenges in cosmology, energy systems, and interstellar exploration.

Cosmology: UED provides a new lens for studying cosmic phenomena, such as the role of the EM field in dark matter and dark energy. This perspective could reshape our understanding of the universe's expansion and structure. By framing the Big Bang as a localized activation of the energy medium, UED inspires alternative models for the universe's origins and evolution.

Renewable Energy: The principles of stable and active energy transitions could guide the development of next-generation renewable energy technologies, such as high-efficiency solar panels or advanced energy storage systems. Understanding the energy medium's dynamics could lead to breakthroughs in harnessing ambient energy, such as tapping into atmospheric electromagnetic fields for sustainable power.

Space Exploration: UED's concept of the energy medium suggests new propulsion methods that interact directly with the field, potentially enabling more efficient interstellar travel. Mapping the energy medium in space could provide navigational aids for spacecraft, leveraging regions of higher or lower energy density for fuel-efficient trajectories. The EM field's role in sustaining cosmic balance could inspire new ways to shield spacecraft and habitats from radiation or gravitational anomalies.

Conclusion

The principles of Unified Energy Dynamics offer a roadmap for future exploration, bridging theoretical insights with experimental and practical applications. From studying photons and the EM field to innovating in energy systems and space travel, UED's implications extend across scientific and technological frontiers. By pursuing these opportunities, humanity can not only validate the profound ideas within UED but also harness its principles to address global challenges and expand our reach into the cosmos. This vision reaffirms UED's potential as a guiding framework for the future of science and exploration.

Appendix B: Mathematical Foundations of UED

Unified Energy Dynamics (UED) provides a comprehensive framework for understanding the universe through the interplay of stable and active energy within the infinite energy pool. The mathematical foundations of UED are essential for rigorously supporting its principles and applications. This appendix introduces key equations, derivations, and models that substantiate the theoretical framework.

1. Wave Equations in the Energy Medium

The energy medium can be modeled as a dynamic field, where wave-like disturbances propagate and carry energy.

General Wave Equation:

$$\nabla^2\psi - \frac{1}{v^2}\frac{\partial^2\psi}{\partial t^2} = 0$$

ψ: Wave function representing the disturbance in the energy medium.

v: Speed of wave propagation, determined by the medium's properties.

This equation models the propagation of energy disturbances, including photons and gravitational waves, within the energy medium.

Electromagnetic Wave in the Medium:

$$c = \frac{1}{\sqrt{\epsilon_0\mu_0}}$$

ϵ_0: Permittivity of the medium (vacuum).

μ_0: Permeability of the medium (vacuum).

This equation connects the medium's constants to the speed of light, reinforcing the UED principle that the medium governs the propagation of electromagnetic waves.

2. Gravitational Wave Interpretations

Gravitational waves, as ripples in spacetime, can be reframed in UED as dynamic disturbances within the stable energy medium.

Wave Energy in Gravitational Systems:

$$E_{\text{wave}} = \frac{1}{2}\rho v^2$$

ρ: Energy density of the medium.

v: Propagation speed of the wave.

This equation captures the energy transported by gravitational waves as they traverse the energy medium.

Metric Perturbations in UED:

$$h_{\mu\nu} = A\cos(k_\alpha x^\alpha - \omega t)$$

$h_{\mu\nu}$: Perturbation of the energy medium's stable state.

k_α: Wave vector.

This models how gravitational waves manifest as periodic disturbances in the energy medium.

3. Entropy and Time Evolution

Entropy's influence on the flow of time is a cornerstone of UED. Time is modeled as a relational property dependent on entropy gradients.

Entropy Change Over Time:

$$\frac{dS}{dt} \geq 0$$

S: Entropy of a system.

t: Time.

This equation formalizes the directional "arrow of time" as entropy increases in an isolated system.

Relation to Energy Transitions:

$$\Delta S = \frac{Q}{T}$$

Q: Heat or energy transferred.

T: Temperature.

Entropy's growth reflects energy transitions within the energy medium, driving time's evolution.

4. Stability and Active Energy Dynamics

The balance between stable and active energy forms the essence of UED.

Energy Conservation in the Medium:

$$E_{\text{total}} = E_{\text{stable}} + E_{\text{active}}$$

E_{stable}: Energy contributing to the medium's coherence.

E_{active}: Energy driving motion and transformations.

Harmonic Oscillator Model:

$$E = \frac{1}{2}kx^2 + \frac{1}{2}mv^2$$

This equation represents the interplay of potential (stable) and kinetic (active) energy within systems governed by UED.

5. Unified Representation of Time, Light Speed, and Entropy

UED integrates time, light speed, and entropy as interconnected phenomena.

Time Dilation in the Medium:

$$\Delta t = \Delta t_0 \sqrt{1 - \frac{v^2}{c^2}}$$

Time dilation reflects the influence of the energy medium on time's flow as relative speeds approach light speed.

Entropy's Role in Light Propagation:

$$v = \frac{c}{1 + \alpha S}$$

α: Proportionality constant linking entropy (S) to variations in light speed (v) in non-uniform media.

Conclusion

The mathematical models and equations presented here provide a rigorous foundation for the Unified Energy Dynamics framework. These tools allow for deeper exploration of the energy medium, wave propagation, entropy's influence on time, and the balance of stable and active energy. Future experimental validations of these equations could further establish UED as a unifying theory of existence.

Appendix C: Stable Energy as the Foundation of the Universe

Unified Energy Dynamics (UED) introduces stable energy as the fundamental substrate that underpins the universe's vastness, balance, and coherence. This concept redefines traditional views, addressing critical challenges in existing frameworks, such as the fate of entropy, the nature of black holes, and the justification for cosmic expansion. By positioning stable energy at the core, UED provides a comprehensive and logical explanation for phenomena that otherwise remain unresolved.

1. Stable Energy and Conservation of Energy

Infinite Reservoir: Stable energy is the infinite and omnipresent foundation of existence, ensuring that energy is neither created nor destroyed but transitions seamlessly between stable and active forms.

Manifestations in Vacuum and Black Holes: Black holes and vacuum regions are not empty voids but dynamic manifestations of stable energy at different equilibrium states. These regions regulate the conservation of energy by transforming active energy into its stable form.

Universal Conservation: The total energy of the universe is conserved through the governance of stable energy, which orchestrates transformations and redistributions across all scales.

2. Black Holes: Purpose and Role

Black holes, within the UED framework, are not material objects but gateways to the stable energy pool, serving as cosmic drains that ensure energy balance.

Energy Recyclers: Black holes absorb all active energy (light, matter, and entropy) and convert it into stable energy, integrating it back into the infinite energy pool. This process preserves energy conservation and prevents entropy from diverging uncontrollably.

Stabilizers: By neutralizing entropy and absorbing structured energy, black holes maintain the universe's dynamic equilibrium. Their role as stabilizers highlights their essential function in balancing energy states and sustaining cosmic harmony.

Purpose: Black holes facilitate the transition of energy from active to stable states, closing the cycles of creation, transformation, and dissolution.

3. Cosmic Expansion Explained

The expansion of the universe is a manifestation of stable energy's dynamic activation into active forms.

Activation of Stable Energy: Cosmic expansion reflects the transformation of stable energy into light, matter, and gravitational waves, enabling the dynamic evolution of the universe.

Balance of Stability and Dynamism: The expansion is guided by stable energy principles, ensuring that it does not devolve into chaos but instead sustains the universe's structure and order.

4. Resolving Challenges in Existing Frameworks

Entropy and the Universe's Fate:

Existing Challenge: Entropy is predicted to increase indefinitely, leading to a "heat death," but its ultimate state remains unclear.

UED Resolution: Entropy increases cyclically and is managed by black holes, which absorb and recycle it into the stable pool, maintaining universal balance.

Light and Black Holes:

Existing Challenge: The fate of light absorbed by black holes is unknown, raising concerns about energy conservation.

UED Resolution: Light transitions from its photon state into stable energy within the energy pool, ensuring conservation and maintaining balance.

Low Temperature in Vacuum:

Existing Challenge: The extreme cold of vacuum space (~2.7 K) is insufficiently explained in the absence of significant energy interactions.

UED Resolution: The dominance of stable energy in vacuum regions explains the absence of material vibrations and low active energy, resulting in near-zero temperatures.

Structure and Balance of the Universe:

Existing Challenge: How does the universe maintain balance and coherence despite entropy and expansion?

UED Resolution: Stable energy provides the underlying framework that governs the universe's structure and ensures balance, with localized activations sustaining temporary dynamics.

5. Stable Energy and the Vast, Balanced Universe

Stable energy is essential to understanding the coherence and balance of the universe.

Conservation of Energy: Stable energy sustains the transformation of energy without loss, particularly in black holes and during cosmic expansion.

Entropy Management: Entropy is absorbed and recycled into the stable energy pool, preventing the universe from descending into chaotic randomness.

Black Holes as Energy Stabilizers: Black holes facilitate energy transitions, acting as cosmic drains and anchors for universal balance.

Cosmic Expansion: Expansion reflects the activation and redistribution of energy while maintaining harmony through stable energy principles.

Low Vacuum Temperature: Stable energy dominates vacuum regions, where the absence of material vibrations results in extremely low temperatures.

Conclusion

Stable energy, as proposed by UED, is the foundation of the universe's coherence, vastness, and balance. It explains the recycling of energy by black holes, the cyclical management of entropy, and the orderly expansion of the cosmos. Black holes, far from being physical entities, serve as gateways to the stable energy pool, ensuring the universe's dynamic equilibrium. Without stable energy, the existence of such a harmonious and self-sustaining universe would not be possible. This framework sustains the law of conservation and resolves fundamental questions about the nature, purpose, and balance of the universe.

Appendix D: Experimental Validations

Unified Energy Dynamics (UED) presents a comprehensive framework that redefines energy interactions across cosmic, quantum, and material scales. To validate its principles, experimental setups and observations are necessary to bridge theoretical insights with measurable phenomena. This appendix outlines potential experiments and observational strategies to support UED's claims, focusing on photon behavior, gravitational waves, and entropy dynamics.

1. Photon Behavior in the Energy Medium

Photons, as active expressions of the electromagnetic field, play a critical role in UED. Experimental setups can explore how photon behavior aligns with UED principles.

Proposed Experiments:

Propagation in Varying Energy Media: Investigate the speed and wavelength of photons under varying conditions of entropy and density. Set up controlled environments with artificial gravitational gradients or entropy variations to study how photons slow or redshift.

Interaction with High-Entropy Zones: Examine how photons behave near regions with artificially induced high entropy, simulating black hole-like conditions.

Hypothesis: Photons lose their discrete identity and integrate into the energy medium under extreme entropy.

Expected Results: Measurable changes in photon speed, energy, or coherence in response to medium density or entropy.

Observational confirmation of photon dissipation in high-entropy zones, aligning with UED's interpretation of light absorption by black holes.

2. Gravitational Waves as Energy Medium Disturbances

Gravitational waves, within UED, are ripples in the stable energy medium caused by dynamic interactions of massive bodies.

Proposed Experiments:

Precision Wave Mapping: Use advanced detectors (e.g., LIGO, Virgo) to map gravitational waves and correlate their intensity with the properties of the energy medium.

Energy Dissipation Rates: Measure how gravitational waves dissipate over vast cosmic distances, providing insights into the medium's stability and dynamic properties.

Behavior Near Black Holes: Observe gravitational wave patterns as they pass near black holes to detect shifts in amplitude or frequency.

Expected Results: Validation of gravitational waves as disturbances in the stable energy medium, with detectable variations in their propagation near high-entropy regions. Insights into the medium's capacity to absorb and redistribute energy dynamically.

3. Entropy's Influence on Energy Dynamics

Entropy plays a central role in UED, governing energy transitions and the flow of time. Experiments can explore how entropy gradients impact observable phenomena.

Proposed Experiments:

Entropy and Light Speed: Test the UED hypothesis that light speed (v) is inversely proportional to entropy:

$v=c/(1+\alpha S)$

Set up high-entropy environments and measure light propagation to validate the relationship.

Time Perception in Entropy Variations: Simulate entropy changes in quantum systems and measure their effects on time-dependent processes.

Gravitational Entropy Studies: Examine entropy levels near massive objects and correlate them with gravitational effects on time and light.

Expected Results: Observational confirmation that entropy influences light speed and time perception, supporting UED's theoretical framework. Validation of entropy as a dynamic factor shaping energy interactions and temporal flow.

4. Observational Strategies for Black Holes

Black holes, as gateways to the stable energy pool, provide natural laboratories for UED validation.

Proposed Observations:

Photon Dissipation at Event Horizons: Observe the behavior of light near black holes to confirm its transition into stable energy.

Entropy Management: Study black hole entropy to validate its role in energy recycling and stabilization.

Cosmic Drainage in Intergalactic Spaces: Search for black hole-like phenomena in intergalactic regions to test UED's hypothesis of their placement as cosmic drains.

Expected Results: Confirmation of black holes as entropy-neutralizing gateways, with no physical matter but observable energy dynamics.

5. Large-Scale Observations: Cosmic Expansion

The universe's expansion is a key prediction of UED, driven by stable energy activation.

Proposed Observations:

Redshift Studies: Correlate redshift data with entropy distributions and stable energy transitions.

Gravitational Wave Background: Map the cosmic gravitational wave background to identify energy redistribution patterns consistent with UED.

Expected Results: Evidence that expansion reflects stable energy activation rather than chaotic inflation, supporting UED's model of balanced cosmic growth.

Conclusion

Experimental validations play a critical role in substantiating Unified Energy Dynamics. By focusing on photon behavior, gravitational waves, entropy dynamics, and black hole observations, these proposed experiments and observations provide a roadmap for bridging UED's theoretical principles with measurable reality. Success in these endeavors would not only validate UED but also revolutionize our understanding of the universe's energy dynamics and balance.

Appendix E: Philosophical and Ethical Implications

Unified Energy Dynamics (UED) bridges the realms of science and philosophy, offering a holistic framework that addresses not only the mechanics of the universe but also its deeper metaphysical and ethical dimensions. By exploring parallels with ancient wisdom and their impact on humanity's worldview, UED redefines our relationship with the cosmos, inspiring a profound sense of interconnectedness, responsibility, and purpose.

1. Metaphysical Parallels in Ancient Wisdom

UED resonates with the timeless principles articulated in philosophical and spiritual texts, bridging ancient metaphors with modern energy dynamics.

Bhagavad Gita: In the *Gita*, *Brahman* is described as the ultimate, unchanging reality underlying all existence, much like UED's stable energy pool. The *Gita's* emphasis on balance—between action (*Karma*) and detachment—mirrors UED's interplay of stability (stable energy) and dynamism (active energy). The cyclical nature of creation and dissolution (*Srishti* and *Pralaya*) aligns with UED's depiction of energy recycling and entropy management.

Upanishads: The *Mandukya Upanishad* describes consciousness as the essence of existence, akin to UED's view of consciousness as energy awareness. The *Spanda* principle in Kashmir Shaivism, which depicts the universe as a pulsation of energy, resonates with UED's depiction of the stable energy medium sustaining dynamic transformations.

Tao Te Ching: Lao Tzu's concept of the *Tao* as the formless, ever-present source of all things parallels the stable energy pool as the infinite substrate of existence. The *Tao's* balance of opposing forces (*yin-yang*) reflects the harmony of stability and dynamism in UED.

2. UED's Impact on Humanity's Worldview

By offering a unifying perspective of existence, UED challenges humanity to reconsider its place in the universe and adopt a more interconnected and harmonious worldview.

Interconnectedness: UED reveals the universe as a seamless web of energy interactions, where every entity, from particles to galaxies, is connected through the infinite energy pool. This perspective fosters a sense of unity and shared purpose, breaking down artificial divisions between humanity and the cosmos.

Ethical Responsibility: Recognizing that all actions ripple through the energy medium, UED encourages humanity to act with greater awareness and care, aligning with universal principles of balance and sustainability. Ethical considerations in energy use, technological advancements, and environmental stewardship are reframed as responsibilities to maintain the equilibrium of the energy pool.

Purpose and Meaning: UED redefines purpose not as domination over nature but as coexistence and contribution to the universal balance. The cyclical nature of energy transitions inspires a sense of renewal and resilience, emphasizing humanity's role in sustaining the harmony of existence.

3. Philosophical Reflections on Balance and Unity

UED's principles echo profound philosophical truths about the nature of balance, unity, and transformation.

The Cycle of Energy: The cyclical interplay of stable and active energy reflects ancient teachings on the impermanence of form and the eternal nature of essence. This understanding fosters humility and reverence for the universe's self-sustaining intelligence.

Humanity as a Participant: Rather than viewing humanity as separate or superior, UED positions it as an integral part of the energy system. This shift encourages collaborative exploration and sustainable innovation aligned with cosmic principles.

Harmony as a Goal: UED challenges humanity to align with the universal rhythm of stability and dynamism, fostering a harmonious relationship with nature and the cosmos.

4. Ethical Implications of UED

The ethical dimensions of UED extend beyond philosophy into practical guidance for humanity's actions.

Sustainable Practices: UED underscores the importance of sustainable energy use, emphasizing recycling, balance, and minimal disruption to the stable energy pool. Innovation in technology and energy systems must align with principles of conservation and equilibrium.

Global Equity: Recognizing the interconnected nature of existence, UED calls for equitable access to resources and energy, ensuring that humanity's progress benefits all.

Exploration with Purpose: Cosmic exploration and scientific advancement are reframed as acts of discovery and alignment with universal truths, rather than exploitation or domination.

Conclusion

Unified Energy Dynamics offers profound philosophical and ethical insights, drawing parallels with ancient wisdom and redefining humanity's relationship with the cosmos. By emphasizing balance, interconnectedness, and purpose, UED inspires a worldview that transcends divisions and fosters harmony with the universe. Its ethical implications guide humanity toward sustainable, equitable, and meaningful progress, ensuring that our actions contribute to the timeless rhythm of existence.

Appendix F: Real-World Applications

Unified Energy Dynamics (UED) not only offers a profound theoretical framework but also has the potential to revolutionize practical applications across various fields, including technology, cosmology, and renewable energy. By leveraging the principles of stable and active energy, UED provides innovative pathways to address some of humanity's most pressing challenges and expand our technological and scientific horizons.

1. Advancing Technology

UED's principles can inspire groundbreaking technological innovations by rethinking energy dynamics and interactions.

Energy-Efficient Systems: Technologies that mimic the stable energy pool's efficiency can minimize energy losses, enhancing power grids and electronic devices. UED principles can guide the development of advanced materials that optimize energy storage and transmission.

Photonics and Communication: The understanding of photons as quanta of the EM field enables advancements in photonic technologies, such as faster optical communication systems and quantum computing. UED insights into the interaction of photons with entropy and the energy medium could improve laser technologies and data transmission efficiency.

AI and Algorithmic Modeling: The interconnectedness emphasized in UED can inspire more holistic algorithms for artificial intelligence, modeling complex systems with greater precision and adaptability.

2. Contributions to Cosmology

UED redefines our understanding of the universe, offering new approaches to exploring cosmic phenomena.

Cosmic Energy Distribution: Insights into stable energy and its transitions can enhance models of cosmic expansion, helping to explain phenomena like dark energy and galactic formation. Observations of energy recycling through black holes can refine our understanding of entropy management in the universe.

Interstellar Propulsion: UED's concept of interacting with the energy medium suggests innovative methods for spacecraft propulsion, reducing reliance on traditional fuel-based systems. Leveraging energy medium properties could enable energy-efficient navigation through gravitational waves or "energy corridors."

Understanding Black Holes: UED's interpretation of black holes as gateways to the stable energy pool provides a new lens for studying their role in cosmic equilibrium, moving beyond the conventional view of singularities.

3. Renewable Energy Innovations

The balance between stable and active energy in UED offers transformative potential for sustainable energy systems.

High-Efficiency Solar Panels: Applying UED principles to optimize photon absorption and conversion into electricity could significantly improve solar panel efficiency. Photonic insights can lead to advanced designs that minimize energy losses and adapt to varying environmental conditions.

Energy Storage Systems: UED-inspired storage solutions, such as advanced batteries or supercapacitors, could enhance energy retention and reduce degradation over time. Mimicking the stability of the energy pool could lead to longer-lasting, safer energy storage systems.

Ambient Energy Harvesting: Exploring the stable energy medium in vacuum or low-entropy environments could enable technologies that tap into ambient electromagnetic fields for power generation. Such innovations could create decentralized, renewable energy sources for remote or underserved regions.

4. Climate and Environmental Applications

UED's emphasis on energy balance can inspire solutions for addressing environmental challenges.

Carbon-Neutral Technologies: Innovations inspired by stable energy principles could reduce carbon emissions in energy production and industrial processes. Technologies that recycle energy more efficiently could reduce resource consumption and waste.

Entropy Management: By understanding entropy as a dynamic factor, UED could inform strategies for minimizing waste energy in ecosystems, industries, and urban environments.

Sustainable Urban Design: UED's emphasis on interconnected systems can guide city planning, integrating energy-efficient infrastructure with natural systems to create sustainable urban environments.

5. Educational and Philosophical Applications

UED's framework extends beyond physical systems, offering tools for intellectual and philosophical growth.

Holistic Education: UED concepts can foster interdisciplinary approaches in education, blending physics, philosophy, and environmental sciences to cultivate systems thinking. Teaching the balance of stability and dynamism could inspire a generation of innovators who prioritize harmony in technological advancements.

Ethical Guidance: The interconnectedness of energy systems can provide a foundation for ethical decision-making in science, technology, and policy.

Conclusion

Unified Energy Dynamics offers immense potential for practical applications in technology, cosmology, and renewable energy. By harnessing the principles of stable and active energy, humanity can develop innovative solutions to modern challenges while aligning progress with the universe's natural balance. These applications not only promise technological advancements but also ensure a sustainable and harmonious relationship with the cosmos.

Glossary of Terms

Here's a comprehensive glossary of key terms used in the book *Unified Energy Dynamics (UED)* to assist readers in understanding complex concepts.

A

Active Energy: The dynamic form of energy responsible for motion, change, and transformation within the energy medium. It interacts with stable energy to drive processes at all scales of existence.

Atomic Transitions: The process by which electrons in an atom move between energy levels, resulting in the emission or absorption of photons.

C

Consciousness: The awareness arising from energy interactions, viewed in UED as a dynamic property of the energy medium that bridges stable and active energy.

Cosmic Harmony: The interconnected balance of forces, motions, and structures in the universe, resulting from the principles of stability and dynamism in the energy pool.

E

Electromagnetic (EM) Field: A universal framework of energy that permeates space, mediates energy exchanges, and sustains the interactions between particles and waves.

Entropy: A measure of disorder within a system, influencing the flow of time and energy interactions. In UED, entropy is a key factor driving the universe toward equilibrium.

Energy Medium: The infinite substrate that underlies all physical and metaphysical phenomena, integrating stable and active energy dynamics.

Energy Pool: Another term for the energy medium, emphasizing its infinite capacity to sustain, transform, and recycle energy.

F

First Disturbance: The initial activation of the energy medium, considered the origin of observable phenomena and energy dynamics in the universe.

L

Light Speed: The constant speed at which light propagates in a vacuum, determined by the energy medium's permittivity and permeability. A critical constant in UED.

P

Permittivity (ϵ_0): A property of the energy medium that determines its ability to allow electric field lines to pass through it, influencing the speed of light.

Permeability (μ_0) : A property of the energy medium that determines its ability to support magnetic field lines, also influencing the speed of light.

Photons: Quanta of electromagnetic energy, serving as messengers that carry energy and information across the universe.

S

Stable Energy: The latent, unchanging foundation of the energy medium. It provides coherence and structure, serving as the counterbalance to active energy.

Spinning and Orbiting Phenomena: The universal patterns of motion seen in atoms and celestial systems, reflecting the harmonious interaction of stability and motion.

T

Time: An emergent property arising from energy interactions and entropy, viewed in UED as a relational phenomenon rather than an intrinsic characteristic of the universe.

U

Unified Energy Dynamics (UED): A theoretical framework that integrates stability and dynamism as the core principles governing the universe, emphasizing the role of an infinite energy medium.

Universal Intelligence: The self-organizing principles evident in nature, reflecting the intrinsic harmony and adaptability of the energy medium.

W

Wave-Particle Duality: The dual behavior of particles like photons and electrons, which exhibit both wave-like and particle-like properties, explained as dynamic expressions of the energy medium.

Closing Reflection

As we conclude this exploration of Unified Energy Dynamics (UED), we are reminded of the profound harmony that underlies existence. UED unveils a universe bound by the interplay of stability and dynamism, where the infinite energy pool sustains creation, transformation, and dissolution. It offers a unifying perspective that bridges the tangible and intangible, the observable and the abstract, illuminating the interconnectedness of all things.

The principles of UED extend beyond theoretical constructs—they inspire a deeper understanding of humanity's place in the cosmos. By recognizing energy as the foundation of existence, UED invites us to see ourselves not as separate entities but as integral participants in a grand, universal rhythm. This awareness fosters a sense of humility, responsibility, and purpose, urging us to align our actions with the balance and sustainability that define the universe.

The implications of UED are vast, spanning science, philosophy, and ethics. It challenges us to think critically about the origins and fate of the universe, the nature of consciousness, and the role of humanity in shaping the future. By embracing the UED framework, we embark on a journey not just to understand the cosmos but to harmonize with its timeless principles.

As we look ahead, let UED inspire curiosity, exploration, and innovation. Let it remind us that the pursuit of knowledge is not merely about answers but about the continuous unfolding of deeper questions. In this eternal dance of stability and dynamism, we find not only the mechanics of existence but also the essence of what it means to be human.

The universe, vast and boundless, is not a mystery to solve but a symphony to experience. Unified Energy Dynamics offers the lens through which we can witness its beauty, understand its intricacies, and honor its infinite possibilities. Let this be a call to embrace the interconnectedness of all things and to live in harmony with the universe's eternal rhythm.

Bibliography/References

This section provides a list of references and sources that have informed and supported the concepts discussed in *Unified Energy Dynamics (UED)*. It includes foundational works in physics, cosmology, and philosophy, along with relevant texts that bridge scientific and metaphysical ideas.

Scientific References

1. **Newton, Isaac.** *Philosophiæ Naturalis Principia Mathematica* (1687).
    - A foundational work on the laws of motion and gravity, providing insights into stability and motion.
2. **Einstein, Albert.** *Relativity: The Special and General Theory* (1916).
    - Introduced the theory of relativity, redefining space, time, and energy dynamics.
3. **Planck, Max.** *On the Law of Distribution of Energy in the Normal Spectrum* (1901).
    - A seminal paper that established quantum theory, forming the basis for understanding energy quanta.
4. **Hawking, Stephen.** *A Brief History of Time* (1988).
    - Explores the nature of black holes, entropy, and the origins of the universe.
5. **Bohr, Niels.** *On the Constitution of Atoms and Molecules* (1913).
    - Advanced the understanding of atomic structure and energy transitions.
6. **Penrose, Roger.** *The Road to Reality: A Complete Guide to the Laws of the Universe* (2004).
    - A comprehensive exploration of the mathematical and physical laws governing the cosmos.
7. **LIGO Scientific Collaboration.** *Observation of Gravitational Waves from a Binary Black Hole Merger* (2016).
    - A landmark paper detailing the detection of gravitational waves, validating aspects of spacetime dynamics.

Philosophical and Metaphysical References

1. **Bhagavad Gita.** Translations by Eknath Easwaran, Swami Sivananda, and others.

- A timeless philosophical text exploring the nature of existence, balance, and purpose.

2. **Mandukya Upanishad.**
   - A foundational work in Vedanta philosophy, describing the nature of consciousness and existence.
3. **Tao Te Ching, Lao Tzu.** Translations by Stephen Mitchell and others.
   - Explores the concept of the *Tao* as the formless essence underlying all things, resonating with UED's principles.
4. **Heisenberg, Werner.** *Physics and Philosophy: The Revolution in Modern Science* (1958).
   - Examines the philosophical implications of quantum mechanics and its relationship to reality.
5. **Capra, Fritjof.** *The Tao of Physics* (1975).
   - Explores parallels between modern physics and Eastern philosophies, bridging science and metaphysics.

Additional Resources

1. **Kip Thorne.** *Black Holes and Time Warps: Einstein's Outrageous Legacy* (1994).
   - An accessible exploration of black holes and their implications for spacetime.
2. **Stephen Hawking and Roger Penrose.** *The Nature of Space and Time* (1996).
   - A dialogue on the fundamental nature of spacetime and its relationship to the universe's energy dynamics.
3. **Einstein, Albert, and Infeld, Leopold.** *The Evolution of Physics* (1938).
   - A historical account of physics' development, tracing its journey from classical mechanics to relativity.
4. **Smolin, Lee.** *The Trouble with Physics* (2006).
   - A critical look at modern physics and its unanswered questions, inspiring innovative frameworks like UED.

Acknowledgment of Inspiration

This bibliography acknowledges not only direct references but also the intellectual heritage and timeless wisdom that have inspired the creation of Unified Energy Dynamics. The fusion of science, philosophy, and metaphysics aims to honor the interconnected pursuit of knowledge across disciplines and cultures.

Author's Note

As I bring this journey of *Unified Energy Dynamics (UED)* to a close, I am reminded of the endless questions that the universe continues to inspire. This work is not a final answer but a beginning—a framework to explore the infinite complexities of existence, bridging the tangible and intangible, the scientific and the philosophical.

The essence of UED lies in its unifying perspective, where stability and dynamism coalesce to form the foundation of all phenomena. This perspective invites us to look beyond isolated observations and fragmented theories, encouraging us to embrace the interconnectedness and harmony that define the cosmos.

A Personal Reflection

Writing this book has been a journey of discovery and reflection. It has reaffirmed my belief that science and philosophy are not separate pursuits but complementary paths to understanding. The beauty of the universe lies not only in its measurable phenomena but also in its ability to inspire wonder, curiosity, and a sense of purpose.

I am deeply grateful for the teachers, storytellers, and scientists whose wisdom and insights have shaped this work. Their contributions remind us that the pursuit of knowledge is a collective endeavor, one that transcends time and space, linking generations of thinkers across cultures.

To my students and readers, you are the reason this book exists. Your questions, curiosity, and eagerness to explore new ideas inspire me to continue asking "why" and "how." My hope is that this work sparks in you the same sense of wonder and inquiry that it has brought to me.

Future Outlook

The ideas presented in UED are only the beginning of a larger conversation. There are countless avenues for exploration, and I invite thinkers, scientists, and seekers from all disciplines to engage with this framework and expand its boundaries:

Scientific Validation:

Experimental investigations into the properties of the energy medium, photon behavior, and entropy dynamics could provide empirical support for UED principles.

Observations of black holes, gravitational waves, and intergalactic phenomena may reveal new insights that align with or challenge this framework.

Philosophical Exploration:

The metaphysical implications of UED, particularly its parallels with ancient philosophies, offer fertile ground for interdisciplinary studies that bridge science and spirituality.

Practical Applications:

Innovations in technology, renewable energy, and cosmology inspired by UED principles could transform our approach to sustainability and exploration.

Education and Collaboration:

UED invites a holistic approach to education, integrating science, philosophy, and ethics to foster a deeper understanding of the universe and our role within it.

A Call to Curiosity

The universe, vast and boundless, remains an enigma, beckoning us to explore its mysteries. UED is not the final word but a stepping stone, a way to frame the questions that have yet to be answered. As humanity continues its journey of discovery, may we do so with humility, respect, and a profound appreciation for the interconnectedness of all things.

This work is an invitation—to think, to question, to explore, and to contribute. Together, we can expand our understanding of the cosmos and our place within it, ensuring that the timeless rhythm of the universe continues to inspire generations to come.

Thank you for joining me on this journey. The story of Unified Energy Dynamics is yours now to continue.

About the Author

SANDEEP CHAVAN, the author of *Unified Energy Dynamics (UED)*, is an engineer, educator, and visionary thinker whose work bridges the domains of science, philosophy, and metaphysics. With a robust background in mechanical engineering and over two decades of experience spanning industry and education, Chavan has dedicated his life to exploring the principles that govern existence, both tangible and intangible.

As an educator, **Sandeep Chavan** has mentored and inspired countless students, instilling in them a passion for inquiry and a curiosity about the natural world. His approach emphasizes interconnected thinking, encouraging learners to see the unity in diverse disciplines. This belief in holistic understanding serves as the foundation for his work in *Unified Energy Dynamics.*

A Vision for Unified Understanding

In *Unified Energy Dynamics (UED)*, **Sandeep Chavan** presents a revolutionary framework that unifies scientific and philosophical perspectives. Drawing on foundational principles from classical physics, quantum mechanics, and ancient wisdom, he challenges conventional paradigms and offers a fresh perspective on the universe as an interconnected system of energy dynamics.

Chavan's ability to synthesize complex concepts into a cohesive narrative is rooted in his dual passion for scientific rigor and metaphysical exploration. With *UED*, he demonstrates how the principles of stability and dynamism govern everything from subatomic particles to cosmic structures, bridging the tangible realities of physics with the timeless truths of philosophical inquiry.

Inspiration and Purpose

Sandeep Chavan's lifelong fascination with the universe is the driving force behind this work. Through *Unified Energy Dynamics*, he seeks to address profound questions about existence:

What is the fundamental nature of energy and its role in the cosmos?

How can humanity reconcile scientific discovery with philosophical wisdom?

What is our place in the vast and balanced universe?

These questions, shaped by Chavan's extensive teaching experience, reflect his commitment to fostering a deeper understanding of the cosmos. His students' curiosity and enthusiasm have been a constant source of inspiration, fueling his exploration and creativity.

A Legacy of Exploration

With *Unified Energy Dynamics*, **Sandeep Chavan** invites readers to join him on a transformative journey of discovery. By challenging established ideas and proposing new pathways for thought, this book seeks to inspire dialogue and collaboration across disciplines. Chavan's ultimate vision is not merely to explain the universe but to celebrate its infinite beauty and harmony, encouraging humanity to live in alignment with its timeless principles.

Through this work, **Sandeep Chavan** offers more than a theoretical framework—he provides a bridge between worlds, a call to curiosity, and a tribute to the enduring pursuit of knowledge. His dedication to understanding the universe's mysteries is a gift to all who share the quest for meaning and truth.

Disclaimer

The content presented in *Unified Energy Dynamics (UED): Decoding the Universe's Infinite Harmony* is the author's interpretation of scientific, philosophical, and metaphysical concepts. While every effort has been made to ensure accuracy, the ideas and frameworks proposed in this book reflect the author's perspective and are intended to inspire thought, exploration, and discussion. They are not to be considered definitive scientific conclusions or universally accepted theories.

This book is a synthesis of interdisciplinary insights, drawing from established scientific principles, philosophical reflections, and speculative exploration. Readers are encouraged to approach the material with an open mind and critical thinking, recognizing that many of the concepts discussed remain subject to further validation, experimentation, and debate within the scientific community.

The author does not claim that the Unified Energy Dynamics framework replaces existing scientific models or theories. Instead, it is presented as a complementary perspective, offering new ways to interpret and understand the interconnectedness of the universe.

The ideas and interpretations expressed in this book should not be used as a basis for professional, financial, or medical decisions. Readers seeking scientific or philosophical advice are encouraged to consult relevant experts or authoritative sources.

By engaging with this book, the reader acknowledges that the purpose of the content is to inspire inquiry, discussion, and exploration rather than to provide definitive answers or professional guidance. The author assumes no responsibility for any interpretations, conclusions, or actions taken by readers based on the material presented herein.

Thank you for approaching this work with curiosity and a spirit of exploration.

Don't miss out!

Visit the website below and you can sign up to receive emails whenever SANDEEP CHAVAN publishes a new book. There's no charge and no obligation.

https://books2read.com/r/B-A-EPGPC-TIJOF

BOOKS 2 READ

Connecting independent readers to independent writers.

Did you love *Unified Energy Dynamics*? Then you should read *Energy in Action: The Soul Demystified*[1] by SANDEEP CHAVAN!

[2]

Energy in Action: The Soul Demystified by **Sandeep Chavan** offers a fresh and transformative perspective on life and existence, redefining the soul as energy in action—a dynamic force confined to the body during life and seamlessly reintegrating with the universal flow upon death. With a unique blend of philosophy, science, and spirituality, this book explores life as a continuous energy loop, governed by natural laws and sustained through the interplay of consciousness, desires, and interactions.

The book begins by demystifying concepts of the soul and consciousness, peeling away layers of mysticism to reveal their roots in universal energy. It redefines life as an energy game, where desires fuel growth, pain teaches resilience, and connections sustain the rhythm of existence. Through practical examples, such as the kindness ripple effect, the interconnectedness of ecosystems, or the role of instincts in decision-making, Sandeep Chavan makes these abstract concepts relatable and applicable to everyday life.

Key themes include balancing instinct and intellect, understanding subconsciousness, and the profound clarity that acceptance brings when navigating life's transience. The book also reimagines the concept of God, shifting focus from divine absolutes to universal energy, inviting readers to

1. https://books2read.com/u/3JwxAQ

2. https://books2read.com/u/3JwxAQ

celebrate life as it is—a vibrant, interconnected system driven by action, interaction, and transformation.

Purpose is highlighted as a subjective construct, created through individual experiences and contributions rather than dictated by universal absolutes. This perspective encourages readers to embrace the mystery of existence, fostering curiosity, presence, and intentional living. With the help of real-world examples and practical tools, the book offers a grounded approach to aligning with the energy loop, finding harmony, and creating meaning in everyday actions and connections.

Energy in Action is a thought-provoking guide that challenges conventional views while remaining accessible to all readers. It emphasizes simplicity, interconnectedness, and the beauty of living authentically. By bridging ancient wisdom with modern understanding, the book serves as both a philosophical exploration and a practical roadmap for anyone seeking clarity, connection, and fulfillment.

If you're ready to redefine your understanding of life and participate fully in the dynamic energy loop, this book is your perfect companion.

Read more at https://www.sandeepjchavan.com/.

Also by SANDEEP CHAVAN

Echoes of War: The Russia-Ukraine Conflict and Its Global Ripples
Shattered Lives
The War Economy
Borders Redrawn
Arms and Allies
Collateral Damage
Justice in the Ashes

The Eternal Sage Chronicles
The Sage and the Sovereign

The Fundamental Theory of Life & Existence
Energy in Action: The Soul Demystified
The Active Mind
Energy in Connection
Death: The Disconnection

Ultra Modern Life Philosophy
The Duality of Laughter and Tears
Compare to Accept

Standalone

It's Not AI, It's AHI - Amplified Human Intelligence
The Decision Paradox
The Four Sapiens
Malicious Script of Indian Polity
The IIT Legacy & Global Impact
The IIT Legacy & Global Impact
You are just a Version of Your Original
Win the Game You Didn't Choose
Always Keep Your Bags Packed
Reveal to Shield - Learn the Game of Facades
Engage Beyond Elections
Culture, Identity & Change: The Evolution of Indian Society
The Triangular Dynamics
Never Trust Without Doubt
If You Can't Change the Country, Leave the Country
Live Fearless with Fear
India: The Investment Magnet
Realignment
Powering a Billion Dreams
Universal Skill
Extension: From Brush to Algorithm
Human Side of Einstein
The Future Knows No Lines
Let's Bring the Generations to the Table
Unified Energy Dynamics

Watch for more at https://www.sandeepjchavan.com/.

www.ingramcontent.com/pod-product-compliance
Lightning Source LLC
LaVergne TN
LVHW080036210926
848481LV00011B/3204
9798227272379